DIGITAL ANIMATION
BIBLE

Digital Animation Bible

Creating Professional Animation with 3ds max, LightWave, and Maya

George Avgerakis

McGraw-Hill
New York   Chicago   San Francisco   Lisbon
London   Madrid   Mexico City   Milan   New Delhi
San Juan   Seoul   Singapore   Sydney   Toronto

The McGraw·Hill Companies

Library of Congress Cataloging-in-Publication Data

Avgerakis, George.
   Digital animation bible / George Avgerakis.
      p. cm.
   ISBN 0-07-141494-0
   1. Computer animation. I. Title.

TR897.7.A95 2003
   006.6'96—dc22   2003059303

P/N 143500-X
PART OF
ISBN 0-07-141494-0

The sponsoring editor for this book was Stephen S. Chapman and the production supervisor was Pamela A. Pelton. It was set in Century Schoolbook by MacAllister Publishing Services. The art director for the cover was Anthony Landi.

Printed and bound by RR Donnelley.

 This book was printed on recycled, acid-free paper containing a minimum of 50% recycled, de-inked fiber.

McGraw-Hill books are available at special quantity discounts to use as premiums and sales promotions, or for use in corporate training programs. For more information, please write to the Director of Special Sales, Professional Publishing, McGraw-Hill, Two Penn Plaza, New York, NY 10121-2298. Or contact your local bookstore.

To Mario Pastorelli C.
The artist's artist and teacher's teacher.
"Pim Pon Pappas"

CONTENTS

Contents

Contents

PREFACE

Introduction

This book is intended for anyone interested in entering the 3-D animation business and pursuing a professional career in computer-generated imagery (CGI). It will give you a great start along two paths: career building and technical skill advancement.

The career path starts you from knowing absolutely nothing and guides you into an internship, then into building a show reel of your animation work, then into finding a paid job (freelance or staff), and finally into starting your own animation studio.

Among the technical skills, the most daunting is learning one or more animation programs. This book helps you choose which program is right for you by surveying three of the most professionally popular programs, **3ds max**, **LightWave**, and **Maya**.

Other technical skills include buying the right equipment, acquiring additional software and "plug-ins," and learning some of the traditional aspects of cell animation and cinema that serve as the foundation for many of the most advanced 3-D programs. Now for some fun.

What Makes This Book Unique, You Ask?

Well, first off, it's not about any one 3-D animation software product. You can buy big, thick books about 3ds max, Maya, and LightWave. They are jammed with really cool tutorials and technical tips. But they don't allow you to compare, on the beginner's level, the working methods and interfaces of each product. This book does. So before you commit yourself to a program for what may be your entire career, you can learn about animation as a career and then pick a program.

Next, none of the 3-D animation books tell you how to get a job. They don't tell you how to edit a good show reel. They don't tell you how to call a studio and get an internship where you can learn on great equipment for free. This book does that.

And there are even some tutorials in this book. But they're the easy ones that let you compare some of the basic tasks of each animation program so that you can get the feel for them. Which brings me to the included CD.

The Free CD

The CD that comes with this book has a working version of Maya and LightWave and at the back of this book, you will find a tear-out coupon to order a free copy of 3ds max. (The reason I didn't include the 3ds max program on the CD is that it was just too big!)

On the CD you'll also find some goodies from other manufacturers. The first of these is FrameCycler from Iridas that lets you play back full-screen animation frames on your computer (a very useful tool, worth over a hundred bucks).

DarkSim, a software company specializing in plug-ins for animation programs, has also generously agreed to include demo versions of their popular DarkTree shader and Simbiont texture application. You can download more files that expand your texture library from DarkSim directly at www.darksim.com.

The CD and the 3ds max coupon, therefore, provide you with all the tools you need to explore the programs that the majority of professional animators use today to create cutting-edge animations for movies, commercials, educational and industrial films, games, and illustrations. Using this book as your guide, and the software on the CD, you should be able to master at least one program well enough to choose your path and jump-start your career.

Finally, this book is mostly about finding yourself as a *total* animator, not just a 3ds max or LightWave or Maya animator. A total animator is a practitioner that doesn't think much about the mechanics of the software, but concentrates on getting great animation done efficiently enough to have a real life. Being total means having a long-range picture in mind—a kind of storyboard of your life—from the now point to way out there in retirement age. How do you want to fill those years? Do you want to work for somebody else or run your own shop? Do you want to work on movies or make games? Do you want to produce the entire animation or do you want to specialize in one aspect, such as modeling?

This book will show you all these choices and more, with useful insights on the pros and cons of each path. Which path will you choose? The time to decide is right now.

Program Differences

You should know that the manufacturers of animation programs do not, as yet, have club meetings every Thursday night to decide what they are going to name the features of their programs. Because of this sad fact, you will find some of the attributes and features of one system being somewhat different from those of another system.

That's a good reason to read this book before you buy one of the leading animation programs; you'll make a better selection knowing what program has what feature.

And if you already bought a system or are in a job working on one, it's still a good idea to read this book, so you can learn what your fellow animators using other programs are doing, the words they are using, and the tools they have.

During the writing of this book I came across many instances where a term or feature in one program was called something else in another program. I started to build a table, which might, one day, serve as a comprehensive cross-reference between 3ds max, LightWave, and Maya. Did I say "comprehensive?" Forget that. I did say "someday." It's really just a start, but a start that has some value, so I'm going to put it right here, at the beginning of the book. Notice that there are a lot of blank fields and missing terms. That's your first homework assignment. Let's get together on this and fill it out. You can find another copy on the CD and on my web site, www.avekta.com/animbible/glossary.doc. Download it and keep it handy. Whenever you find a new translation, or a correction of one of mine, fill it into the document in a highlighted color and email it to me. I'll continue to update the glossary on the web site and who knows, in a few years, we might have something worthy of the term "comprehensive!"

Best wishes to all the readers, especially to those who so kindly supported my first book and especially to those who wrote me emails and had such nice things to say about the work. It is those letters that kept me going through this book!

GEORGE AVGERAKIS
New York, NY

ACKNOWLEDGMENTS

There are no accidents. Even something as inconsequential as a paperback book, a book that was not considered important enough to be produced hard bound, takes an enormous amount of effort to bring into being. A great portion of that effort comes from people who do not share in the glory of having their name on the cover. The author owes a debt of gratitude to these people, and hopefully the readers who benefit from their generosity will indulge me this brief space to enumerate their contributions.

Representatives of the companies who produce 3-D animation software gave generously of their time, contributed artwork, advice and even software so that you might choose to become a practitioner of the craft they invented. Chuck Baker and Bill Vaughn from NewTek provided software and guidance on LightWave. Michael Stamler, Danielle LaMothe, and Lei Lei Sun from Alias|Wavefront assisted me with Maya. Kevin G. Clark and Kevin Baribalt from discreet coached me through the learning curve on 3ds max.

Kevin also helped provide the cover art, "Cute Girl," by a young artist from Bordeaux, France, named Olivier "Reiv" Ponsonnet. Kevin showed me this stunning piece of work on a web site, www.cgtalk.com, managed by Leonard Teo from Australia, who was also an enormous help in getting this book into print. CGTalk is a great place to meet CG artists from around the world and to show your work and look at the work of other talented artists.

This is my second book for McGraw-Hill. My relationship with this publisher was a result of the generous mentorship of Brian McKernan, editor of *Digital Cinema Magazine*, for whom I wrote many magazine articles during his tenure as editor of V*ideography Magazine*. Brian introduced me to McGraw-Hill editor-in-chief, Steve Chapman, who has, with great personal attention, patiently extracted each of the pages you now hold.

To everyone at MacAllister Publishing Services, I owe continued gratitude for assuring that my grammar, syntax, and basic good taste meet the standards of the civilized world. Without their assistance, you would know why my high school English teachers predicted I would never publish a single word.

Finally, I would also like to express my deepest gratitude to master animator, Arnold Gallardo, author of *3D Lighting: History, Concepts, and Techniques*.[1] Arnold meticulously read my manuscript *two times* to detect

---

[1]Charles River Media, Book and CD-ROM edition (2000), ISBN: 1584500387.

omissions and errors and to clear up passages that were not as clearly written as they could have been. As an animation producer, my perspective on the industry is different from that of a working animator and Arnold's viewpoint kept the focus on this book well balanced between the art of animation and the management of animation.

Regardless of the efforts of all the people whom I have thanked here, I fear there are still some errors lurking in these pages. The vast complexity of animation science, the ever-changing parameters of three enormous programs, and my own human inadequacies lead me to believe there was, perhaps, another year of tweaking left to do here! Ah, but deadlines must be met—even one which was kindly extended three months as in the case of this book. So, if you should find an error or serious omission, you have my sincere apologies. Furthermore, if you would care to assist in perfecting the *Digital Animation Bible*, you can send your comments to me directly by email: george@avekta.com. I will do my best to incorporate your corrections in future editions.

Glossary from 3ds max Point of View

3ds max	LightWave	Maya Instance Attribute Editor
Aimpoint		
Animation	Layout	Animation
Atmospheric		
Bones	Bones	Bones/Skeletons
Character Icon?	Null Object	
Collisions		
Constraint	Range	Constraint
Curve Point	Node	Node
Default Shader = Phong	Default Shader =	Default Shader = Blinn
Dynamics Simulators		
Edge Visibility Threshold	Smoothing Angle	All Soft Angle
Edge Visibility Threshold	Edge Opacity	All Soft Angles
Fog		
Function Curves	Motion Graphs	
Global Illumination—Light Tracer		
Global Illumination—Radiosity		
Graph Editor	Envelope	Graph Editor
Gravity		
IK Goal	NullGoal	End Effector (or Gnomon)
Key Button in default	Create Key frame for Specific Object	Set Key (S key)
Key Button with Key Filters or Track View selections	Create Key Frame for All Objects	
Materials	Surfaces	Materials
Materials Editor	Surface Editor	Materials Editor
Menu Bar (over hidden) Tab Panel	Panel Tabs	
Modeling	Modeling	Modeling
Modifiers		
Morphing		
Motion Controllers		
Move	Move	Move
Named Selection Set	Numeric Input	Channel Box
Opacity	Transparency	
Option Box		
Origin, also 0.0,0	Origin	
Parametric (bend, twist)		
Particle Systems		
Pivot Point	Pivot Pint	Pivot Point
Properties Dialog		
Rotate	Rotate	Rotate
Rotate (view)	Rotate (View)	Tumble (View)
Scale	Scale	Scale
Scene	Scene	Workspace
Secondary Motion (Flex)		
Selection Sets/Subsets/Objects within Sets	Parent—Child	Parent—Child
Self-Illumination	Luminance	Self-Illumination
Shaded/Wireframe Toggle = F3 key	Shaded/Wireframe Toggle =	Shaded/Wireframe Toggle = 5 key/4 key
Soft Bodies		Cloth
Special Effects		
Specular Level	Specularity	Shininess
Tab Panel	Menus	Menu Sets
Trackbar	Timeline	
Viewports	Views	Views

ABOUT THE AUTHOR

George Avgerakis was born in Trenton, New Jersey, and educated in the public school system of Ewing Township. As an undergraduate at the University of Maryland, he taught filmmaking under the guidance of fine artist/educator, Henry Niese. Upon graduation, Mr. Avgerakis attended the London Film School, where he studied under film director, Mike Leigh.

Mr. Avgerakis began his professional career in London, as a post-production scriptwriter, later moving to Princeton, New Jersey where he worked as a freelance film editor and director. He has lectured at London University, City University of New York, and at various seminars in the United States and abroad. After brief engagements as an art director, TV commercial producer, and corporate video manager, he cofounded Avekta Productions in 1982, establishing a studio in midtown Manhattan, for which he serves as V.P. Creative Director to this day.

Mr. Avgerakis has produced work for such clients as AOL-Time Warner, Phillips Petroleum, Forbes, Akzo Nobel, Brink's, Hewlett-Packard, Pfizer, and the American Museum of Natural History. He produced the first co-production between a corporation and the Public Broadcasting Network, and the first infomercial for Russia after the fall of the Soviet Union.

A contributing editor to *Videography and Digital Cinema* magazine, Mr. Avgerakis has published an average of 5,000 words monthly since he began writing technical articles in 1990.

Mr. Avgerakis lives in Yonkers, New York, with his wife, Maria. He is the father of Stephanie, a cadet at the United States Military Academy at West Point and a son, Alexander, a student at the Pratt Institute.

Mr. Avgerakis welcomes letters from readers and can be reached by email at **george@avekta.com** or at this book's web site, **www.bizbible.biz**.

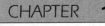

Your First Day on the Job

What is it like to be an animator? Is it a profitable profession, allowing you the time and money to enjoy life while basking in the glow of artistic admiration from friends and strangers alike? Is it fun to get up and go to work every morning, to feel hungry for lunch and find that it's 3 o'clock in the afternoon, and to regret having to call it a day while others are watching the clock tick slowly toward five? Is it a possible career and not just some dream job that one in a million people get to do?

Well, yes. It is all these things—sort of. As I write these words, after more than 15 years in one small corner of the animation business, only two things in that first paragraph need to be clarified. The first, and biggest, is that you really have to want to be an animator for the job to be a good one. You have to have a passion for the craft and a talent for the art. The second, a minor point really, is the "time and money" part. You can certainly make good money. Salaries for full-time, experienced animators can be excellent, but you may not get the time to enjoy the money. Animation is a consuming profession. As one Los Angeles veteran says, "It eats you up alive."

The Passion

It gets easier and easier to get hooked on animation. The computer software you'll use to create the most sophisticated effects and character scenes gets easier to learn and cheaper to buy. You might get a chance as early as high school to play with a "lite" version of a program or even the real, full-blown release. Or maybe you will get to visit an animation studio during some vacation tour to Orlando or New York.

Someone may sit you down at a workstation and show you how to bring a sphere or cube into existence. Simple. You know this stuff from geometry, only it wasn't so easy to understand. Then the mentor adds the dimension of time by placing the object in one place at Frame 1 and another place at Frame 30. For fun, the mentor asks you to make some change to the object in Frame 30. Thinking, perhaps, that you can fool the system, you screw around with the controls and the object ends up like Diana Ross, upside down, and inside out.

Nonplussed, the mentor smiles and hits a button. In seconds, the computer creates 28 frames between Frames 1 and 30, adding subtle, successive changes to the original object in Frame 1. By Frame 28, the computer has—miracle of miracles—perfectly imitated your outrageous precocity, *just one small increment shy* of your Frame 30 results. With another button press, the mentor shows you how the machine has animated the results of your demented challenge. It is like nothing you've ever seen before.

Is the machine teasing you? Is it like those video games you know so well, the ones that let you win just enough before causing you to lose so embarrassingly that you have to keep shoving quarter after quarter into its hungry maw so that you can rescue your dwindling self-esteem and prove that brains are better than circuitry? Yes, it is like that, but better. It is free. No quarters. Better yet, they say if you're good enough, you can get paid to do this.

"Once more," you plead, taking the mouse in your hands, but the mentor is pressed for time. You have to move on. The tour proceeds, and you are hooked.

Maybe, if you are lucky, you have a chance to go beyond the first "free hit" on the animation computer. Maybe you get the summer camp course or the gifted and talented program elective, or you have a relative in the business (we all should have one) who gives you some time on the system. There you delve deeper into the mysteries of plastic space and time. If so, you learn that you can find no release, no end to the joy of limitless creation. The addiction is deeply rooted now. For you, no help can be offered, except perhaps this book, which charts out the path to a successful, full-time, day-job career. Hello, my name is George Avgerakis and I'm an animaholic.

The Pay

When computer animation first appeared, it made its impact first in the U.S. art community, which is centered in New York City. Guys like Nam June Paik, the Whitney brothers, Ed Emschwiller, and Walter Wright employed various analog devices to create unusual, colorful effects on television screens that were a sensation in both traditional and avant-garde galleries throughout Manhattan. One of Paik's ideas was to manipulate magnets around the back of a color picture tube to distort the electron guns that draw TV images. Paik and a Japanese engineer named Shuya Abe created the first video synthesizer in 1970, using it to produce several broadcast art shows on WGBH in Boston and WNET in New York.

The creative directors at New York advertising agencies, always on the prowl for a new look, quickly discovered the new electronic art form. In the early '70s, computers were just starting to make their capabilities known to this same group of hungry eyes, but computers weren't yet able to emulate the effects that were being shown at the Museum of Modern Art or the Whitney Museum. Analog effects still ruled.

Seeing an opportunity to crack the lucrative TV commercial market, two New York video mavens, Bill Etra and Steve Rutt, joined forces to improve on Paik's idea and market a useful production tool that could be used to make commercial effects. Etra and Rutt invented the Rutt-Etra Video Synthesizer, an analog device with a sophisticated collection of controls that enabled an operator to precisely create colorized video distortions within controlled envelopes of time. The Rutt-Etra Video Synthesizer used a black and white monitor with controllable beam deflectors to warp the screen image according to the demands of a creative director's imagination.

Artwork would be put on a table, shot with a black and white camera, fed into the synthesizer, and then warped. By pressing a button, the warp could be reconfigured predictably. One of the most successful applications of the Rutt-Etra Video Synthesizer was a logo for Merrill Lynch where a stock performance chart morphed into Merrill Lynch's charging bull logo. That assignment and others like it propelled Etra, Rutt, and a number of other creative individuals who purchased the Rutt-Etra Video Synthesizer, and other analog tools like it, into a cadre of successful TV commercial producers who were, erroneously at first, called "computer animators."

In those days, the average television commercial budget ran up to about $60,000. This included the director, camera crew, camera equipment, sets, lighting, actors, and editing. To compare this with current TV commercial budgets, multiply by about 10.

Steve Rutt and his associates would come into an advertising agency, blow the client's nylons off, ask whatever they wanted for a 30-second "computer-animated" spot, and get it. Imagine. All by yourself, with a pile of video equipment in an office measuring 10 by 10, working for maybe 30 hours, you could crank out a national TV spot that you could sell for half a million (2003) dollars. Well, it wasn't always that great. Sometimes the agencies just needed a special-effect bit to slug into an otherwise filmed commercial.

"Okay, we'll animate the text over your picture for, hmm . . . " Steve would say, estimating the job at three or four hours work with a testy art director and a fidgety producer breathing down his neck, "$10,000" (the equivalent of a hundred grand today). Hands would shake; Steve didn't even bother with contracts. "Just sign the work order when we'll start."

Whenever something new arrives in the marketplace, it usually costs more than it should. Needless to say, once a new thing becomes an older thing, the prices drop. Other guys get into the game, price wars and tech wars ignite, and the whole enterprise boils down to two factors: who you know and how good you are. That's fair. Knowing the right people to call is a well-recognized human value, just as artistic competence. The trouble is, the two factors rarely exist in the right proportions in the same person.

So how much can you make as an animator? It depends on your mix of the two factors: sales ability and artistic competence. At the low end, a good industrial animator, cranking out logos and charts for a corporate facility, along with the ancillary applications like web sites and CDs, can make a salary of $30,000 to $50,000 a year with nice benefits.

A sales rep for the same kind of work can do a bit better. A rep at this level, working for a facility that sells to corporations, would get about $40,000 as a *draw salary* against 10 percent commissions. This means that you collect a salary with benefits, but you have to generate $400,000 of sales per year to justify that salary. If you sell more, you make more, based on 10 percent of whatever you sell each year over $400,000. If you make less, you keep your draw, but you risk getting *downsized* (fired). Essentially, your income is unlimited or, rather, limited only by the size of your market, the economy, and your sales skills.

I hear you saying, "Hey, I want to be an animator, not a salesman." Right. But what if you combine the animator's salary of, say, $40,000 to a sales commission bonus of, say, a modest $20,000? Sound interesting? Good. Now if I show you some simple techniques that would not take you too far off the animation path yet yield you 50 percent more take-home pay, would you walk away from that? I didn't think so. Read on.

Moving up from there, an animator working for a TV commercial production company in New York, like Click 3X, Curious Pictures, or R/GA, could make from $50,000 to $80,000 with royalties or bonuses as a senior artist with a client following. Looking at the West Coast, a similar animator, working at a *computer-generated imagery* (CGI) or visual effects company, cranking out elements for feature films, television series, or entire animated productions for both the small and large screen, would enjoy the same salary range, though perhaps a bit more overtime and stress.

Animators in the lucrative gaming industry command both a salary and royalty expectations as their products go from development into long-term lifetimes in both set-top box products and in multiplayer Internet environments. In such cases, the royalties may, at times, exceed the salary, especially if the animator has an ownership or equity position with the production company. Of course, the lure of royalties also allows some producers to offer significantly lower salaries, creating a risk/reward profile, which may only attract the young and foolish.

The longer forms of an animated television production or a feature film come from the corporate studio, rather than the animation shop dedicated to shorter projects. Although television and film work is more stressful, in my opinion, a bit more money can be made, and a deeper sense of family exists among the animators. A long-running episodic television program, such as the 1990's hits *Space Above and Beyond* and *Babylon 5*, provided many young animators with opportunities to grow, supervise, and even direct episodes. The longer a show runs, the more congenial the work atmosphere becomes and the more attractive the job is from a career-building standpoint.

Your Own Animation Studio

More than teaching you how to make animations, this book is a guide to building your career. Everyone reading this book has one significant advantage over anyone that tried to become an animator 5 or 10 years ago, and it pertains to Moore's Law. In 1965, Gordon Moore, cofounder of Intel, established a theory based on the number of transistors per square inch on *integrated circuit* (IC) boards, noting that the density doubled every year since the IC's invention. The chips on an IC are composed of vast electronic circuits. The amount of circuits that can be reduced into one chip, increases as chip manufacturing techniques get better. What was once an entire room of circuits, costing millions, now fits in a chip costing a few cents, mounted on a board, costing $100. Next year, that board is a chip costing 10¢ and so on.

Moore concluded that this trend would continue for the foreseeable future, although the "doubling time" has recently settled down to about 18 months.

Of course, you could infer a corollary to Moore's law, based on the tendency of computer owners to always want the fastest computers. This corollary would add, "And as the new double-speed model enters the marketplace, the preceding model's price will drop proportionately." Stated quite simply, Moore's Law means that almost every two years it becomes twice as easy for you to own all the tools necessary to establish and run your own animation studio.

As of this writing, one of the most common configurations for a high-end workstation consists of a dual Pentium III processor machine. Each processor runs at the maximum speed of 1.33 *gigahertz* (GHz, meaning 1 billion), or 330 billion cycles per second), and they are installed with up to 4GB of *Synchronous Dynamic Random Access Memory* (SDRAM). Users willing to sacrifice the dual-processor capability and opt for the fastest single processor (known as a *uniprocessor* configuration) can, as of this date, acquire up to 3.06 GHz of processing speed (that's 3 billion, 60 million cycles per second). Users can then employ the latest Pentium IV, which does not as yet support dual processor installation. (See www.intel.com for the latest information on processor capacity and compatibility.) For uniprocessor applications, many animators swear by *Advanced Micro Devices,* Inc. AMD Corporation's *Athlon MP* processor, which, operating currently at 2.13 GHz, is comparable to the Intel Xeon 2.8 GHz processor and sells for considerably less. (See AMD's web site benchmark results at www.amd.com or specifically at www.amd.com/us-en/Processors/ProductInformation/0,,30_118_6291_2003,15590,00.html).

For the writing of this book, I have been the fortunate recipient of three excellent computers you could well afford and acquire for your own studio. In some cases, the optional equipment added to these computers represents more than 75 percent of the cost. That means you can acquire these tools at a much lower cost than the following totals and then upgrade them by adding options as your business grows. If you have such a plan, consider the basic, single processor workstation with one monitor, a utility-grade video card, and two small hard drives.

The first of the computers used in writing this book is a Hewlett-Packard (formerly Compaq) EVO W8000. It is equipped with two 2.8 GHz Intel Xeon processors, 1GB of *Double Data Rate* (DDR) RAM, and a 3Dlabs Wildcat III 6110 AGP Pro video card (with 384MB of online memory). It also has a 36GB Seagate Cheetah system drive, a 181GB Seagate Barracuda drive for storing media and sequences, and a Compaq TFT1825 flat-panel monitor. This configuration would retail for about $10,000. I also installed NewTek's LightWave 7.5 animation software ($1,300) on this computer. You can reduce the Hewlett Packard EVO W8000, from $10,000 to less than $2,000 and still have plenty of computer power to get started in animation production if you choose fewer options.

The second computer is an IBM InteliStation Z Pro, equipped with two Intel Xeon Two 8 GHz processors, 1GB of DDR RAM, and a Nvidia Quadro4 video card with 128MB of RAM (see Figure 1-1). It also has an 18GB IBM Ultra 160 *Small Computer Systems Interface* (SCSI) hard drive for programs and a 73GB IBM Ultra160 SCSI for media. The monitor is an IBM P77 17-inch flat-screen monitor (the package price would be approximately $8,000). On this computer, I installed Discreet's 3ds max 5 animation software ($3,495), along with a wide assortment of Digimation plug-in bundles, including Particle ($700), Character Animation ($1,000), Materials ($700), Nature ($950), and Special Effects ($900).

The third computer is a Compaq EVO W8000 workstation with a 3.06 GHz Intel Pentium IV processor, 1GB of DDR RAM, a 36GB Seagate

Figure 1-1
The IBM
IntelliStation Z-Pro
ensemble

Cheetah system drive, a 181GB Seagate Barracuda drive for media, and a Nvidia GeForce4 video card with 128MB of RAM. This setup uses a Compaq CV7500, 17-inch flat-screen monitor (the package price was approximately $4,700). On this computer I installed Alias|Wavefront's Maya 5 Complete animation software ($1,999).

In each of these systems I also installed several other software tools indispensable in animation work, such as Adobe Photoshop, Adobe Illustrator, Adobe After Effects, and eyeon Fusion. Although your choices of platforms and software may vary considerably from what my choices were, and even my choices will change as new computers and hardware come onto the market, none of these systems, ready to work and earn you money, costs more than $10,000. Adding on the required peripheral gear should run you no more than another $5,000 to $8,000, depending on your needs, making the complete investment at or around $15,000. This is certainly a fraction of what it cost to start a computer animation company just five years ago. Chapter 3, "The Equipment You Will Need," will go into more detail about a professional animation studio's requirements, including strategies you can use to obtain the right system for your business goals.

Each of these workstations is capable of allowing an animator to create characters and complex 3-D objects, assemble them in vast scenes of geographical and biological detail, and move the characters and objects realistically in lengthy scenarios. In short, these workstations are minifactories you can purchase for not much more than the cost of a compact automobile, although the trip you take on them is endless and highly profitable. Such power for such a small cost has not always been the case.

One Man's Story

My mother is a traditional woman who grew up during the Great Depression. People like that always tell their children to get a good *full-time* job with a pension and good health benefits. If they're like my mother, they add on things like "And find a nice girl, get married, settle down, have kids, and buy a nice house with a white picket fence." My mother would always suggest I go up the road from my house to a company called Squibb in Princeton, New Jersey. Even back then, before it merged with Bristol Myers to become one of the largest pharmaceutical companies in the world, Squibb represented my mother's dream place to work. The vast, manicured lawn, punctuated by ponds, across which, rare aquatic birds swam, was my mother's ideal of security. Any company that could afford ponds would never fire a good kid like her son. Mothers don't want to see their kids grow up to be freelancers or even work in the turbulent world of media. They'd rather we grow up to be cowboys.

I got my first animation computer in 1983. In those days, the best animation computer was a Bosch FGS 4000, which cost several hundred thousand dollars. I went to a *Society of Motion Picture and Television Engineer*s (SMPTE) conference to see it demonstrated and was awed to see the operator move around within the cockpit of a space ship and then push out through the wall to show the exterior. I couldn't get this effect out of my mind.

A month or so later, probably influenced over the years by my mother's nagging, I managed to get an interview to show my video productions samples at Squibb. The in-house video manager had been to see the Bosch FGS 4000 too, and we got to talking about the kinds of animation this monstrous computer could render. Suddenly, the client offered me a chance to bid on the design of the corporation's logo in 3-D animation. I had never sold an animation project before, but I told the client I could deliver what he wanted. I went home and called Bosch to find out who owned an FGS 4000 in my neighborhood.

It turned out that another New Jersey corporation, AT&T, owned one. I called them and asked if they would rent it to me or use my storyboard to make my logo project. A *storyboard* is a pencil sketch, kind of like a comic book, that allows an animation artist to inexpensively

(continued)

One Man's Story *(continued)*

describe an animation. AT&T and I struck a deal. I sat down for two days and sketched out two animation storyboards, one very high-tech and one traditional.

I took these boards back to the client at Squibb and he really liked them. He said he'd call me back in a few days to tell me which one the company liked. In two days, he called me and told me Squibb wanted both animations: the high-tech one for the United States and the traditional one for Britain. Generously, he offered me twice the money I quoted as the "either one" price, although I would have given him a discount if he'd asked.

The high-tech logo would have to be created on the Bosch, but the traditional video could be created on a Via Video machine that had just been introduced to the market, selling for a mere $45,000. It was my idea to buy the machine with the profits from the job, but I still needed another $25,000. I called my parents to borrow the money.

"Hey, Ma. I got a job with Squibb."

My mother was speechless, but only for a few seconds. "Full time with benefits?"

"No, no. It's a project assignment."

I could hear her sigh with great disappointment.

"But, Ma, it pays more than I made in six months of last year."

"Oh?"

"And I get to design Squibb's corporate logo. The whole identity of the corporation is in my hands."

"No health plan?"

"No, Ma, but now I can afford to buy one. That is if you and Dad can help me with something."

"George, you should find yourself a nice girl—"

"Mom, I'm already married."

After some critical evaluation of my business plan, my parents loaned me some money and I bought a Via Video computer . I should also add that the current president, Ronald Reagan, had added a powerful incentive to investing in your own business. The government was allowing a 10 percent tax credit (not a deduction—a *credit*) on any new equipment purchased by a business or individual. Say what you will about Mr. Reagan, but that credit was like a ladder dropped down from the ruling class with an invitation to come on up. It doesn't exist anymore.

To give you some comparison between this investment and something you can relate to, two years earlier I had bought my first home, a four-story, handyman brownstone in downtown Brooklyn for $52,000 with $5,000 cash down. Can you imagine having to spend the equivalent of a house on your first 3-D animation computer?

Today that brownstone is worth well over a million dollars (I don't own it anymore unfortunately), but the Via Video is now doing service as the fender of a Hummer or an I-beam. I sold it for $1,000 in 1995 and was darn lucky to get that. So was the computer a bad investment? No. The Via Video paid for itself in a year and helped me win several clients that I would not have won otherwise. I paid back my parents way ahead of schedule and all was good. So here's the lesson for that story: Buy a small apartment building for long-term financial security and never sell it except to buy another apartment building. Buy a computer to do animation and make fast money, but only pay as much as you can recoup completely in one year. More on this issue later.

So now I had an animation computer and a show reel with two examples of my work. What was the next step? Computer animation was a new thing. So far, only the wealthiest clients—TV commercial agencies and feature film producers—could afford it. With one phone call, I could get appointments with the top ad agency's creative directors, but I was always hit with a Catch-22. I didn't have enough examples of my work on the show reel. I had to spend some of my own money and time making *speculative* commercials. This would normally be doing work for free. Not me. Not with a loan to pay off. I had to generate work and money fast. Where to go?

Sitting around one day listening to a business program on the radio, I heard a guy talking about how he set up a new kind of office coffee machine. Nobody wanted to buy his coffee machine because they didn't think it would work. So he took a chance and offered a special deal where if the customer wasn't satisfied, he wouldn't have to pay a cent for the machine. The idea worked for him and none of his customers asked for their money back.

I thought, I can do that. If the client asked for his money back, I would only be back to making a speculative project anyway. As long as

(continued)

One Man's Story *(continued)*

the work I was getting improved the show reel, I was making headway. I ran this idea past my wife and partner, Tica, who never gets blinded by the kinds of fantasies that often delude Hollywood types like her husband. She clamped down on the idea and channeled it well.

"Here's what you do. Make this offer only to the corporate clients. Don't send it to the agencies. They won't respect your offer and even if you succeed, they will always think you're a price beggar."

Man, was she right. Years before, when I'd been an assistant producer at Grey Advertising, all the producers had all the commercial directors pigeonholed into budget and subject categories. There were kid directors, food directors, comedy directors, and, alas, low-budget directors. The low-budget directors got a lot of cheap jobs that nobody else would touch, but they never got a shot at doing a quality spot, and even if they did, their reputation would never allow them to get a decent budget. TV commercial production is a small community with slow turnover. Whatever reputation you make in the beginning will most likely follow you throughout your career.

In the end, I never got into commercials. The money-back guarantee idea worked just fine and I ended up with a string of corporate clients who were delighted to learn about a new computer that could give them TV-commercial-quality animation at an affordable price. Because my computer cost 25 percent of what the large animation studios were paying, I could carve myself a nice niche, make good money, and never have to spend a cent courting the TV commercial market.

The lesson of this part of the story: Find a niche and serve it with pride.

Selling Your Talent

So now you have an idea of what it is like to get into this business as an owner-producer. If you like to sell, that is, pick up a phone and pitch your creativity to a stranger, you have an excellent chance of making a good living.

If you don't like to sell, you can consider hiring a salesperson or hooking up with a *representative*. These two choices presume that you already have a stunning show reel of successful projects and perhaps a group of loyal clients willing to follow you (kind of like a hairstylist moving from salon to salon). You aren't likely to be able to afford a salesperson on your staff or to attract a representative as a novice.

If you don't like to sell and you're still rather new at the trade, you'll have to go to work for someone else. To get that work, you're going to have to (sorry) sell yourself. You're going to take your reel, however weak or strong it is, and pitch it to a producer, visual effects supervisor, or animation house owner. To get that pitch, you're going to have to do some research (to get the names of the people you need to meet), make some calls, and send some mail. You may be thinking, "I can't sell!" Yes, you can, and I'll show you how. And it will be easy.

If you don't know it yet, I hope you realize that the better you can sell, the more money you will make. Consider this: If you work for a producer or animation house, chances are the amount you get paid will be less than half of what your boss charges his clients for your work. That's a simple formula. Most companies, to be profitable, have to *mark up* your costs. If you get $50 per hour, the client will pay more than $100 per hour to your boss. Your boss uses his share, $50 or more, to pay his expenses to run the shop, make sales calls, and have a bit left over to buy a yacht and a house in Malibu. That's how it works.

Now if you have a representative, you will pay this representative anywhere from 10 to 15 percent of your gross billings. The representative will usually expect an allowance for lunches and other forms of entertainment to get clients to meet with him or her on your behalf. If you're really good, you can hire a full-time representative, pay him or her a salary and benefits, and thereby reduce the overall expense—*if* you consistently sell your work and have a regular income. But then you're running your own production company, aren't you? If you run your own production company, you make the most money per hour of creative effort.

Categories of Animation Business

Chances are, you will succeed in any one or more of the seven basic animation categories:

- Hollywood feature films
- TV commercials
- Televised entertainment

- Games
- Home videos
- Business communications
- E-media

Hollywood Feature Films   Hollywood feature films need not actually be produced in Los Angeles, California, but it really helps if you are located somewhere near this geo-creative nexus. Several excellent animation studios working on feature films exist as far away as Santa Barbara, Seattle, and even Massachusetts. Obviously, the benefit of a global Internet marketplace makes locating your animation business in Bombay, India, practical (and I would bet that the savvy Hollywood producer will one day discover the cost benefits of global sourcing). However, feature film production is still a lunchie, backslapping, *meshpuka*[1] kind of business where personal contact represents a major part of the deal. It's a little hard to schmooze over a T-1 line.

Cracking into a gig doing 3-D animation for feature films is kind of like cracking into feature films as a director. You have to be really persistent, concentrate exclusively on this market, and be a bit lucky. If you have a really good show reel, particularly with regard to character animation and *compositing,* you might get a break. You'll need to find a new director or producer who is looking to get some good, cheap CGI into a film and needs to find a low-budget supplier to make the budget. It's like the music video business where you find an unsigned band and get to be their video person. They make it? So do you. Short of that leap, you can send your reel to existing shops that already work the feature film market and hope their needs and your reel match up.

TV Commercials   TV commercials are another hard market to crack, but easier than feature films. As the commercials get higher in budget, the market gets harder to crack. If you live in a second- or third-tier market (that is, a city that is not New York, Chicago, or Los Angeles), you can pitch your reel to advertising agencies and commercial producers. If your reel is a starter, you should still be able to get some work doing cable commercials, but if you have a stick of dynamite with a sizzling fuse, head for a top-tier market, hire a rep, and go for broke.

---

[1]Yiddish, the unofficial language of Hollywood, gives us this powerful word, which means family.

The top markets are dominated by the big advertising agencies that have an established custom of seeing new talent. You basically call producers or, better yet, their assistants, send your reel, and pester them a bit until they see it. It's a hit or miss proposition. Having a rep with existing client contacts helps enormously. Reps come in all flavors, from the ones who have their own businesses, to reps you hire yourself on a full-time basis. Maybe, on second thought, it's a good idea to get a start in a small town, build a good reel, and work your way to one of the three coasts.

Televised Entertainment Televised entertainment includes TV shows that use 3-D animation and composited effects as part of their weekly fare. Again, this is primarily a Los Angeles market, but nothing is stopping you from creating your own show in Smalltown, USA, pitching it to a network, and getting a sale. Mike Judge did that with *Beavis and Butt-head,* although he used primitive, 2-D cell animation instead of sophisticated 3-D computer techniques.

The cost of producing a half-hour or hour-long pilot using 3-D animation is rather small compared to making a filmed presentation. You're biggest expense will be the voices. Mike Judge did several of his own!

Cable television, especially the public access channels, is a great place to get started. You can test your production techniques, audience response, and enjoyment for the process with little risks.

Games Game devices, from arcades to set-top boxes (and computers in between), continually increase their capabilities to emulate real 3-D, photorealistic action. The market for 3-D animation in the gaming community is always growing and highly competitive. Skills that will place you at the head of the line include character and creature development, achieving photorealistic playback with minimum memory allocation, understanding moving camera dynamics, and, of course, good teamwork.

You may consider designing your own game and using it to penetrate the market. Certain equipment requirements, such as a motion capture system, may be beyond your reach, but your creative concept, supported by less sophisticated choreography, may be adequate to earn you a significant production contract, or at least a great job.

Home Videos Home Videos are another outlet for creative 3-D animation. If you decide to make your own televised entertainment, you can extend the selling opportunities of your work by sending it to home video distributors.

You can also go to your local video rental house, look over the special interest videos section, and collect the names of key producers. These enterprising businesspeople are always looking for a way to spice up their productions, and if you can price a package of logos, graphics, special effects, and other eye candy for their productions, you will have found a friend and a long-lasting client.

Business Communications Business communications is the largest market for 3-D animation, especially videos and DVDs that are made to explain the arcane intricacies of medicine and high technology. Here your market is composed of producers, marketing directors, human resource managers, venture capital entrepreneurs, and training directors, all of whom have a constant need to have their communications embellished by animation and graphics.

Many of these clients have liberal criteria regarding the quality and complexity of the animations they purchase. If you are a beginner, and your reel is not yet replete with the most original, cutting-edge work, you may still be able to make sales in this category. Your client may not have had the opportunity to see some sophisticated work or does not understand the difference between the reels of a beginner and a seasoned pro. This limited window of opportunity should not be exploited by laziness. If you are lucky to get a client after only a few months' practice in 3-D, don't sit on your laurels and take your good fortune for granted. Keep pushing yourself and your skills.

E-Media E-media presentations that appear over the Internet represent another large and growing market, which temporarily offers advantages to the beginner. Because most e-media is streamed to the viewer at a comparatively low bandwidth, the complexity of the animation it can play is severely limited. Beginners, whose work is limited by their level of knowledge, can exploit a medium whose resolution is limited by its bandwidth if the beginner is aware of how these two limitations overlap.

Low bandwidth means fewer colors and shading, fewer frames, simpler morphing, and less detail than full-bandwidth imagery. If you plan to get your start in this category, concentrate your initial skill set on achieving good results within the limitations of the medium and expand your skill set as the medium's bandwidth increases. You just may get in on the ground floor of something big.

Your First Job: What's It Like?

What's it like if you're a new animator, just hired, and you're walking into your first day on the job? Here are two scenarios that illustrate first-day experiences for a small business communications shop and a large Los Angeles animation studio. I've taken the liberty of converting some real-life interviews of three entry-level animators into a narrative story that each animator has reviewed for accuracy. I hope you enjoy reading their stories this way. In the case of Humble Lantern, I've changed the names in order to respond to the studio owner's request or privacy.

Small Corporate Shop

Frank Dexter came to work for Humble Lantern as an unpaid intern with about 12 months' experience of working on his own in 3ds max. Prior to this, he'd graduated from a good Eastern with a degree in art and design, a fluency in French, and other suitable intellectual pursuits. Frank got the intern position after making several calls to Humble Lantern, offering his services for free in turn for equal time with a well-equipped computer so that he could perfect his show reel.

At the time of Frank's arrival, his shop had no specific work for him, so he was shown a workstation, a collection of tutorial books, and documentation that referred to the software on Frank's computer. He was then left pretty much alone. Frank's officemate was another animator, Suzanne Drake, who had been working for Humble Lantern as a full-time freelancer for about two years.

A full-time freelancer is a person, paid by the day or week, who receives a higher day rate than a salaried employee but does not receive any benefits, such as medical coverage or retirement. Like any freelancer, Suzanne can be hired for projects as the workload increases or decreases. Many people prefer to work freelance because they do not want the benefits and prefer the freedom of choosing their workplace and schedule on a day-to-day or, in Suzanne's case, a month-to-month basis.

For a month or so, Frank kept pretty much to himself, chatting occasionally with Suzanne as he started to build an impressive array of 3-D characters. Some were humanoid, whereas others were fanciful animals.

During this time, the owner of Humble Lantern, Jim Brandt, frequently looked over Frank's shoulder and offered whatever suggestions and assistance Frank needed. Eventually, Brandt became impressed with Frank's work and started mentioning his new animator's work to Humble Lantern's clients.

Meanwhile, Suzanne had been offered a lucrative full-time position at MTV. Although her new job would be limited in its creative scope because she would be animating one of the web site features, the prestige, the much higher salary, the benefits, and the chance for advancement led Suzanne to take the offer and tender a two-week notice of resignation with Magic Lantern. As soon as Jim Brandt learned about Suzanne's departure date, he invited Frank Dexter into his office to discuss a new work arrangement.

Instead of being an unpaid intern, Frank was offered a new challenge. Brandt had recently taken some of Frank's characters to a long-term client. The client was a major home products manufacturer looking for a humorous animation character to symbolize its new line of cookware marketed to men. Brandt had pitched the creation of a brawny Irishman with an iron skillet in his fist. The client liked the idea and was willing to pay Magic Lantern an exploration fee to see what kind of prototype characters could be developed.

Frank possessed a great skill at cartooning and in a few weeks had pencil-sketched several dozen ideas for the new character, each with a distinctly different look and attitude, yet each fitting the client's desire for a brawny Irish cook. Eventually, the client selected a composite of three drawings and commissioned Magic Lantern to create the character in **Maya**, which Frank had already grown accustomed to using during his internship. Within a month, the client accepted Frank's designs, earning a second sizable commission from the client.

Brandt, seeing a major talent forming in his studio, generously shared the income with Frank, and for a period of time, Frank's income soared to levels he had never expected he'd earn after just three months of interning. The character was eventually developed into a series of three commercials, which ran on national television, thus earning both Frank and Magic Lantern a significant entry for their show reels.

Frank Dexter remained with Magic Lantern for another year, completing a total of six commercials, enough to build an impressive show reel. Frank would have stayed much longer at Magic Lantern had not a recession hit the industry that summer, leading Brandt to reduce Frank's paid hours of work to a bare minimum. Hungry for more challenging assignments, Frank asked Brandt for a letter of recommendation, which Brandt was happy to compose from glowing compliments. In less than two months, Frank was offered a job at a larger animation studio that secured a long-term animation contract with a television network.

Frank's new job was as a technical director of animation on a new Saturday morning children's series. His new duties include managing five animators and two interns, the specification of new equipment purchases, equipment maintenance, and creative direction. Although Frank had made about $35,000 in his first year at Magic Lantern, his new salary started at $50,000 with two weeks' paid vacation, as well as medical and retirement benefits. Eight months after Frank started his new job, he was given an 8 percent raise.

A Midsize Disney Subcontractor

Antonio "Tony" Bryant was a senior at the Pratt Institute when the school's internship placement office sent him to an interview at Cartoon Pizza in Manhattan. Cartoon Pizza creates animation for a wide variety of broadcast clients, one of them being the Disney Channel.

When asked to show his portfolio, Tony opened up a book that was divided into several useful categories of commercial art. It included typography, logo designs, charcoal life drawings to demonstrate traditional skills, a corporate stationery assignment, a web site design, and a print campaign to demonstrate his ability to take on a limited commercial task. Then there were the characters. The second half of Tony's portfolio included a menagerie of characters ranging from cartoon elves to intricate robots. They inhabited scenes ranging from jungle bogs to chiseled castles. They were shown as pencil sketches, wireframes, orthogonal diagrams, and finished renders.

It was clear to Tony's prospective employer that he was interviewing a significant talent, but that is not a rare occurrence. What was rare was that the talent was demonstrated on both the brilliant and the mundane sorts of projects, and each sparkled with enthusiasm. Tony was permitted to intern for Cartoon Pizza's standard six-week tenure, ending with the New Year holiday.

Tony's duties were to execute revisions of hand-drawn work: settings, scenes, and characters. The work involved taking scans of the original version and, using Adobe Photoshop, electronically pasting in the corrections, which were given to Tony as hand-sketched inserts. "The trick," recalls Tony, "was getting the corrections to match seamlessly with the original. It wasn't easy to do."

Notice that even when faced with a rather laborious, mundane task, the aggressive intern, interested in proving himself, made the tasks interesting and challenging. "I got quite good at the work," said Tony, "and finished early. So I went around to the senior artists' desks and asked them if they needed any help. Sometimes it was going out for a coffee and sometimes it was a challenging art assignment. One time I was asked to design a character for a Disney Channel pilot."

Tony submitted dozens of sketches for the character but didn't neglect his regular job or the artists who were still asking for a double latté with Sweet & Low. Eventually, the New Year holidays arrived and Tony packed up his gear and returned to Brooklyn to search for more work. Tony was a December graduate from Pratt and was now officially looking for a job.

Just after the holidays, Tony got an invitation to a party at Cartoon Pizza. Everyone at the party kept remarking how much he was missed around the shop. The next morning he received a call from the supervisor of his internship, offering him a regular freelance position.

As of this writing, Tony's original character design has been accepted by Disney for a new series. His name will be in the credits, and although it looks like he's solid for work, he still cleans up other people's artwork and fetches a coffee or two. At Pratt, Tony's nickname was "Ghost," because he suddenly disappeared and reappeared amongst his friends without notice. All that's changed. Tony sums it up: "Being the kind of person people enjoy having around on a reliable basis is the key to success." (See Figure 1-2.)

A Major-League Hollywood Studio

This story goes back a bit in time, because I want to show you what can happen a few years after you get your first job. John Gross was going nowhere as an assistant at a swimming pool company in Minnesota where he'd worked for 10 years. A friend had bought a computer and was making some money doing computer graphics, so as John puts it, "I borrowed some money from my parents, with the idea that I would give myself five years to get into computer graphics. Well, in about a year and a half I was working part-time for a company called Alphavideo in Minnesota. After another six months, they hired me full time."

John became the in-house expert on LightWave, back when NewTek was just introducing the software (ancient times in computer lore, about 1991!). But he yearned for greater horizons and eventually heard about a computer

Figure 1-2
Animator Tony
Bryant at his
workstation at
Cartoon Pizza in
New York City

animation studio called Amblin Imaging that was being set up by Steven Spielberg to handle all the computer generated imagery (CGI) for a TV series to be known as SeaQuest. John sent his reel to Amblin and was offered a job. Eventually, he became one of the two top animator-managers of the shop, working directly with Spielberg. In addition, John's ability to squeeze the maximum efficiency from both computers and artists created an excess capacity at Amblin. Excess capacity means that a shop has more computers and labor than it needs to be profitable. Obviously, if the capacity can be used for more work, the studio can make even more profit. So Amblin started bidding on outside work, such as the space effects for *Star Trek Voyager*.

Eventually, Spielberg decided to shut down Amblin when he formed his current partnership, DreamWorks SKG. For a while, Gross and Spielberg discussed being partners in an independent animation studio that John would helm, but eventually John went off on his own, and today runs Eden FX, one of the most successful Hollywood animation shops. On a regular basis you can see John Gross's name rolling up the credits of major films and television shows (see Figure 1-3).

Figure 1-3
John Gross, right, being interviewed by the author for a documentary during the shutdown of Steven Spielberg's Amblin Imaging, just prior to John starting his own animation studio

The point of all these stories is this. If you decide what it is you want to do and where you want to do it, you can get there if you invest a little bit of money and a lot of work. Once there, if you continue to work hard at it, you can go really far. The sky, or at least a virtual copy of it, is the limit.

How This Book Works

When I sat down to start this book, I didn't want to open it up with a lot of boring stuff about how this book was organized or what you could expect to get out of it. That's always a cool glass of ice down your back when you're fired up to get into the meat and potatoes of a new career. So I started you off with some tasty nibbles of what your career as an animator could be like. Now that you've bit that off and are chewing it through your imagination, here's that cold beverage to wash it down.

This book combines two essential knowledge paths to being a successful animator. First, it tells you about the basics of animation, which involve the creation of a storyboard, objects, scenes, and characters. Secondly, it tells you about the business of animation that involve getting an internship, making a show reel, getting a paying job, and starting your own business. I decided to write a chapter for each step on both paths. To make the book fun to read, I decided to alternate the chapters between animation craft and business science. To add icing to the cake, I interviewed several professional animators who allowed me to put their own experiences in this book.

The result, I hope, is a fun and educational book to read, with practical software applications that enable you to get out of the starting blocks and toward reaching your goals.

Boldface Type

In order to help you learn the terminology of the animation business, I've put all the new words in **boldface type**. These words may also be found in the index. Therefore, if you search for a word in the index, you can go to the page listed and easily find it by looking for the boldface type.

Fictitious Names

For various reasons, some of the stories in this book use fictitious names for people and companies. Any resemblance between these names and real people is coincidental. For the sake of brevity, I have not noted which names are fictitious and which actual, so you may amuse yourself with conjecture or Internet searches for possible contacts.

The Most Popular Software

Prior to starting this book, my publisher, McGraw Hill and I did a survey to find out what was the most popular animation software. Our results revealed three animation products: NewTek's *LightWave*, discreet's *3ds max*, and *Alias | Wavefront's Maya*, all of whom assisted in the creation of this book.

My goal is not to produce a shootout competition between these three companies and their offerings, but to provide my readers with a comprehensive overview of the tools. Consequently, rather than accentuating the differences between the products, their commendable features, and regrettable shortcomings, my effort is to promote a general appreciation for and understanding of all of them. My wish is that you become an expert in, or at least become familiar with, all three.

My experience as a professional animator and creative director for a small corporate media facility I cofounded in New York City in 1982 has given me a broad perspective on both the animation business specifically and the e-media business in general. As a contributing editor for two major trade journals, I have also had the opportunity to review just about every new animation and graphic program on the market since 1991.

During these years of feast and famine, I've seen many programs and computer systems come and go. Although nothing is certain in our world, and life is even more precarious in the world of computer graphic technology, the three animation products we have selected for inclusion in this book have several important aspects that rank them as first rate. More importantly, they possess the highest probability of being in existence for many more years. It's a shame to spend months learning an animation program only to see it disappear from the market. Nothing is guaranteed in the software business, but the risk of these three programs evaporating from the animation marketplace is very low.

Although I focus on just three animation software programs, many others are on the market. Some have unique advantages in design or price, making them worthy of exploration. In addition, ancillary graphic tools assist the animator in completing productions. Certain programs are necessary tools in the animation process. These include **compositing** (the combining of animated and live-action frames), **photo manipulation** (cutting, pasting, and redrawing scanned art), **illustration, font management** (using text and logo designs in animations), **standards conversion** (changing the various video and frame formats to conform to your needs), and **picture compression** (reducing the data density of animations so they can run on systems with narrower bandwidth than your original creation). Several of these programs will be mentioned in this book, and more are invented every day to make your life easier. I encourage you to search the net for the key words that I've boldfaced above, read the trade press and visit conventions where new software is introduced.

The Workflow Process

Regardless of the type of animation you are contemplating, its complexity, audience, deadline, or budget, the same general workflow takes place and the same eight steps apply. Keep in mind that not all steps are required for every project and that, unless you are a studio owner, you may not be required to execute all the steps in your specific job category.

1. Pitching the project (Chapter 1)

2. Contracting and billing (Chapter 1)

3. Storyboard (Chapter 2, "From Concept to Screen: The Workflow Process")

4. Objects, scenes, and characters (Chapter 5, "Objects and Surfaces," and Chapter 7, "Rigging and Animating Characters")

5. Motions (Chapter 7)

6. Testing (Chapter 1)

7. Rendering (Chapter 6, "Let There Be (Rendered) Light")

8. Getting paid (Chapter 1)

Because of the manner in which I've edited this book, the chronological order of these steps do not correspond to the chapter order, but here is a brief description of each step.

Pitching the Project

Unless you are creating the animation for yourself, you will have to explain your animation to someone else. This someone might be your roommate, your boss, or your client. Usually, this other person has some decision-making role in the process of creating the animation. If the animation is for your friend, and no payment is involved, your friend may respect the value of your labor contribution and be tentative with her comments and requests. If, however, the animation is for a paying client or your boss, it is your responsibility to initiate and maintain an effective flow of information about the animation.

Many clients and some bosses have no idea what goes into the process of animation. Worse, many well-heeled decision makers have gross misunderstandings of what an animator can and cannot do.

One client of mine brought me a videotape of a famous celebrity presenting the front of a box containing an electric toothbrush package. He wanted me to rotate the box in 3-D space to reveal the artwork on the back of the package. I asked the client for a photo of the package's back with the idea that I would create a 3-D model of the package, position it in the scene, and attach animated hands that would replace the celebrity's actual hands. Using these elements, I could, with some labor, make the actress appear to rotate the package to show the back. The client shook his head with disdain. "I don't have the back of the package," he explained. "That's my point. I need to know what is on the back without going back to the manufacturer. Can't you just rotate it with your computers?"

After questioning the client a bit more, I ascertained that he had recently seen a movie called **_Enemy of the State._** It featured a scene in which a team of high-tech government graphics engineers, working with a security surveillance tape, reading the name off the side of a shopping bag that was not facing the camera. Using their computer skills, the technicians in the movie rotated the bag in 3-D space to reveal the logo printed on the back of the bag. The client wanted to pay me "any amount of money" to execute the same effect with my computer equipment.

I don't know what this client's background was or why he couldn't simply call the package designer. Maybe he'd messed up and didn't want to expose his mistake. Maybe he was a corporate spy pretending to be an advertising executive. This would not have been the first time a practitioner of covert ops knocked on my door. The trouble was I couldn't take this client's money because he had such a gross misconception of animation capabilities. You simply cannot rotate a 2-D object in a video to reveal something you would not otherwise be able to see, nor do I think it will ever be possible.

This little story illustrates just one example about how you must communicate your limitations with a client. Additionally, you must communicate these limitations _before_ they become a problematic issue. Knowing your limitations is one of the most difficult pieces of knowledge to acquire as a professional. The reason for this is partly ego (it's not easy to admit you can't do something) and partly knowledge itself (you may not _know_ what you _don't_ know until it's too late).

In addition to limitations, good communications establish a common understanding between client and animator. When a client says, "I want a cute lion," what she pictures in her mind might be totally different from what you imagine. Often, the client may not have anything pictorial in mind, just something "cute." Your job as a communicator should be to establish what, in pictorial terms, the client considers cute. Draw a few faces. "Do big round eyes mean cute?" you ask her, as you draw a lion face with big doe eyes. She frowns and says, "No, that's not quite it." So you try square eyes, add big lashes with a pair of Ben Franklin glasses, and see how she reacts. This is all part of the communications process.

Maybe it takes a week or more of sketching, for which you might not get immediately paid, to establish a common language or a visual syntax between yourself and the client. Is it worth it? Maybe not for a one-time $2,000 animation for a client who will never hire you again. But can you be sure a client is ever a one-timer? Not in my 20 years of experience.

Pitching the project therefore begins with the establishment of a common understanding with the client. This understanding consists of creating a link between the client's needs and what you are capable of creating. For instance, if a client wants to make the logo of her product well known to preschoolers, you can propose a Saturday morning cartoon animation that captivates attention and then transforms into the logo. Such a solution would be an **animation concept**.

Once you and the client agree on a concept, the next stage of the production process begins, usually under the control of a contract, which you (or your supervisor) draw up with the client.

Contracting and Billing

It is sometimes difficult to draw the line between the free or **speculative** (i.e. "on spec") work an animator must do to obtain a contract and the paid work that comes after the contract is in force. Often an animator or animation firm will venture up to the animation concept stage and no further before locking the client down to an agreement. Other times the animator will proceed to a storyboard stage and *then* a contract is signed. I've chosen to place the contracting phase before the storyboard stage here, but you could easily argue for the storyboard phase to come next. It all depends on how much you want to do on spec.

Once you decide how much work you will do on spec, it is time to create a contract for your client that will determine what you will do and how much you will get paid. Although many contracts have common elements, such as the names of the parties involved and in which state or province the contract is considered to have originated, animation contracts are somewhat unique in that they must carefully describe an end product, which is to some extent indescribable. You can, for instance, include a storyboard, sketches, and specifications of length and format, but animations change as they are created. You must be careful to create a contract that allows you the necessary flexibility to create, yet that constricts both parties into a defined limit of duties and payments.

Duties are usually expressed as a list of particular assets, which the animator creates according to a schedule of measurable events, or **milestones**. It is customary to receive a first partial payment from the client, in advance of work commencement, upon signing the contract. Usually, this amount is a third of the total anticipated fee. At some point in the process, a second third payment is received, and at the conclusion of the project, upon approval and acceptance by the client, the final payment is received.

Storyboard

The storyboard is the blueprint for the animation. I strongly urge any animator to lock down as many creative aspects of the project as possible in a storyboard before venturing to the computer. The storyboard contains drawings, measuring at least 3 by 4 inches (7.5 by 10 centimeters) with accompanying text. The text describes the animation, any sound effects or music, and any dialog or narration.

Within the animation studio, the storyboard serves to organize the creative aspects of the production. It establishes the level of detail required, the list of objects required in each scene, and the degree of texturing, effects, and other complexities that will be needed. Outside the studio, the animator uses the storyboard as a means of presenting his or her ideas to a client before initiating the expensive process of computer design.

Objects, Scenes, and Characters

All the work you do on a computer animation program takes place in the *graphical user interface* (**GUI**, pronounced "gooey") of the program (see Figures 1-4, 1-5, and 1-6). Although each animation program has a different GUI, over the years, as animation programs have matured, they have, like automobiles, tended to agree on some common design elements. The most common element is the concept of cameras' **viewports** or *points of view* (**POVs**) through which the program "looks" as it composes the images you direct. Furthermore, each POV is directed at a three-dimensional representation of space in which you may place objects that appear in your animation. Obviously, this concept arises from the cinematic arts. Many film concepts and terminology are employed in animation software.

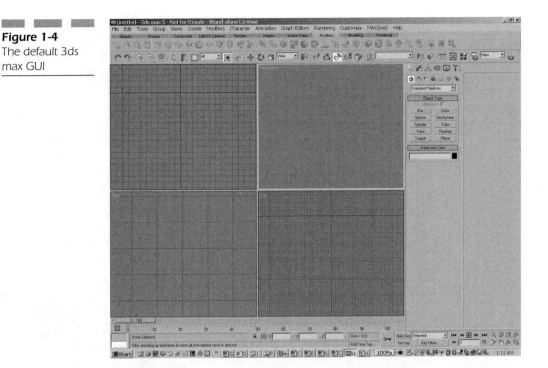

Figure 1-4

The default 3ds max GUI

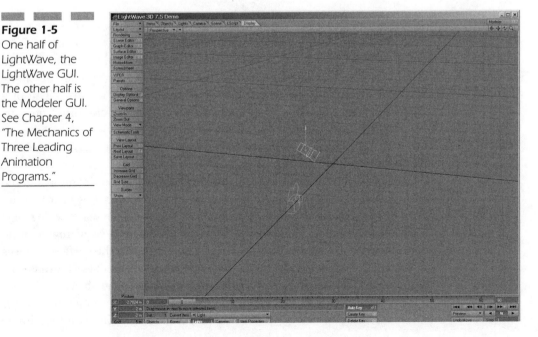

Figure 1-5

One half of LightWave, the LightWave GUI. The other half is the Modeler GUI. See Chapter 4, "The Mechanics of Three Leading Animation Programs."

Figure 1-6
The default GUI
for Maya

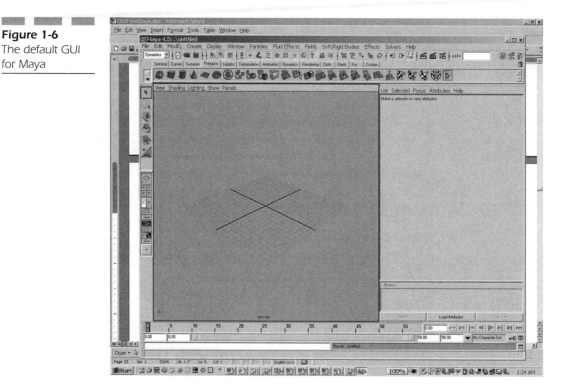

Another common element is the use of Cartesian planes to express graduated space in three dimensions (X for horizontal, Y for vertical, and Z for
depth). ("Egad!" I hear you scream. "We suffered those in high school math
class!" Yes, this is one of those times your teacher was right when he or she
said, "Pay attention! You're going to need this stuff some day.") The programs also offer the capability to view the work you are creating from various viewports, such as through a camera, top-bottom view, right-left side
view, and front-back view. These attributes derive from the architectural
and engineering professions.

Usually, the first step in creating an animation is to acquire or design the
key **objects**. These objects can be as simple as a text-based logo or as complex as a photorealistic human character. Objects may possess a wide range
of **attributes**, including **shape, color, surface texture, luminosity** (interior light source), **reflectance** (the amount of light that the surface bounces
back into the scene), **specularity** (the shininess of the object), **mass** (the
weight of the object), and **resilience** (the hardness of the object).

Objects may also be **articulated**, or bent at specific hinge points. For
instance, a human object, in order to bend an arm, will require a hinge

somewhere within the arm. Consequently, limitations on the movement may imitate the limitations of a real human arm. Changes in the object, resulting from an articulation, such as the bulging of a bicep, can also be specified and linked to the articulation such that the changes are activated automatically whenever the motion is initiated.

Articulations and their associated changes on an object can also be gathered together into *forward kinetic* (**FK**) and *inverse kinetic* (**IK**) chains or **hierarchies**, whereby one articulation may cause a collection of further articulations. FK and IK chains are extremely useful in developing complex character movements. A frequent example of hierarchical control would be movements of a character's hand through space. If the the hand is pushed by the bones of the shoulder and arm, this motion is said to be forward kinetic. Inverse kinetics is where a character moves the tip of one finger and the rest of the fingers, the hand, wrist, elbow, and shoulder all comply proportionately with realistic results (see Figure 1-7). An easy way to remember the difference between FK and IK is to remember the phrase "push forward, pull inverse." Both methods deal with the same objects, but achieve their results from opposite applications of force.

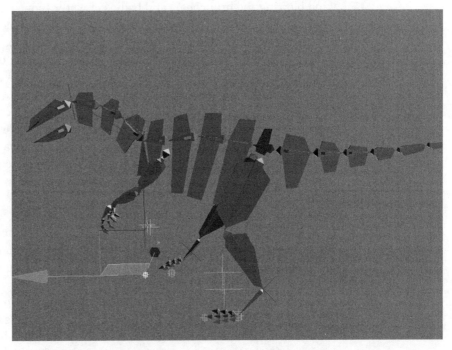

Figure 1-7
An example of an IK chain in the control structure of a character. The rank of joints, articulations, and shape changes can be expressed as a text outline (see the Parameter Wiring dialog box in 3ds max) or by color-coded references on the objects themselves. This dinosaur's right leg is being extended forward by pulling the last bone in the leg chain.

In addition to objects, an animation usually requires the design of a **scene**. If we use a theatrical analogy to understand the relationship between objects and scenes, objects would be actors and scenes would be sets or locations. Scenes can contain a **backdrop, geography, props, lights,** and **special effects** such as **atmosphere, fog,** and **particle systems** or **streams**.

A backdrop is an economical way of creating scene details, which require little time and computer power. Imagine you are watching a high school play of *Man of La Mancha*. Instead of building a real windmill, the school's art department has painted the scene of a windmill, blue sky, green fields, and pebbled road to hang on the back wall of the stage. You can use the same technique in animation by acquiring a photo or graphic design and placing it on a two-dimensional plane within your scene. By carefully facing the camera directly at the backdrop and placing all objects appropriately between the backdrop and the camera, you can often obviate the need for designing a complex 3-D set.

Geography may be introduced to your scene to create the vastness of space (by adding planets and stars) or the hills and rivers of a quaint country homestead. More complex than a backdrop, geographical elements might include textures (rock or grass), clouds (do they move?), trees (is the wind blowing the branches?) and bodies of water (how deep, how clear?).

Props are objects placed in the scene that will interact with characters. A tree, in *Shrek*, for instance, might go from being a geographical element to a prop if Shrek were to rip it out of the ground. A set element, which is acted upon by an object or character in the animation, is a **prop**. Props need more detail and features than simple geographic elements that do not move or that move according to a simple, cyclical, or random control parameter.

Lights are extremely important in animation. Theoretically, without a light in the scene, the picture would always be a black frame. In fact, when an animator does a test render and gets black, the problem is most likely a lighting failure. Computer-generated lighting closely follows the science and art of cinematic lighting. Lights may be named for their cinematic equivalent, such as a **spotlight** (a narrow, intense beam), a **fill light** (a soft, wide-angled beam) or an **ambient light** (no specific source but it affects the entire scene). In some software products, an object is given internal lighting (luminosity) by virtue of its surface attributes (as in LightWave) or by installing a light within the object.

Lights have attributes that can be set numerically in the programs to adjust the affect of the lights in your scene. These include brightness, color, the **focal length** or **cone angle** (the width of the light beam), the **falloff or dropoff** (the hardness or softness of the edge of the light), and 3-D space coordinates (X, Y, and Z increments). Like all objects in a scene, lights have the ability to vary their parameters over time by having their attributes defined by a keyframe. In some programs (LightWave), the operator can set a view through a light's POV.

Additional lighting effects borrow ideas from the movies, such as **lens flares**, **halos**, and **glows**. These are cinematographic devices and probably would not exist in computer animation if photography and cinema had not preceded the computer.

Motions

The word, "animation," implies that a fourth element is being included: time. Without the passage of time, animation could not exist because every scene would be a still frame. Time in animation is defined by a sequence of individual frames, where each frame represents a distinct increment of time. For instance, most *standard definition* (**SD**) videos employ 30 **frames per second**. An animation that will be output in SD video has a basic time increment of 1/30 of a second, and each frame you animate represents that increment of time. If you want to create one second of SD video animation, you will set your software to 30 frames per second and create 30 frames.

Animation programs display the passage of time. The length of the **timeline**, graduated in frames, represents the length of the animation. A **cursor**, which can be moved forward and backward across the length of the timeline, indicates the current frame being displayed on the **playback monitor.** Typical *video tape recorder* (**VTR**) controls for stop, play, rewind, fast-forward, and frame-by-frame shuttling enable the animator to move across the timeline.

All changes to an object's attributes—its position, color, lighting, surface texture, virtually any definable element—can be assigned specific points in time. If, for instance, an animation is 90 frames in duration (3 seconds at 30 frames per second), the object can be assigned a specific list of attributes in frame 1. Frame 1 then becomes a **key frame** for that object. The animator then moves the timeline cursor to frame 90 charges all of the attributes which were defined in frame 1, and creates a new key frame for the selected object. The animation program is capable of creating a smooth transition from frame 1 to frame 90, adjusting each attribute proportionately between frames. This process is known as **interpolating**. It results in the object's movement from its starting position in frame 1 to its final position in frame 90. By assigning a further keyframe for the object in frame 120, the object could be taken to a second position and so on. By using multiple keyframes on multiple objects, complex motions can be created.

Testing

Once you have created an element of an animation, you will want to test it before you ask the computer to produce final still frames (a process called **rendering**), incorporate the **sequence** of frames into the final animation. It is not unusual for a complex animation to require a half an hour per frame to render, sometimes a lot more. A talented operator cannot waste time waiting for a sequence to render so he or she can check the results and find the mistakes. Frequent testing avoids mistakes before the final render. Such mistakes, if overlooked, may become extremely laborious to correct later.

The first level of testing is to simply examine your work onscreen. Depending on your software and computer capabilities, objects can be viewed in different ways. They can appear to be composed of a grid of horizontal and vertical lines, known as **wireframes**. Objects can also take the form of simple polygon **proxy shapes**; a head might be represented as a box, for instance. Your work can also be viewed as **monochromatic** or one-color shapes, which allow you to examine all qualities except the finished color. Objects can also be displayed using any combination of the above methods.

The ability to view a scene in nearly finished quality is enhanced by your computer's video display card which may provide sophisticated processing and memory circuits, while they do their more mundane job of transferring your computer's images to the monitor. Some cards have sufficient memory and processing capabilities to enable you to see a nearly finished version of your objects as still images or even fully animated sequences. Such emulations are called **Open GL** renderings. Although they offer the animator an approximation of the finished work, their level of quality depends on your computer's video display card, and they are not complete or entirely accurate representations of the final rendering. They are, however, a good way to test your work *before* rendering. It's also wise to bring the client into the studio (or to send the client digital video clips or still frames of the animation) to examine your work as it progresses through object and character testing.

Another testing method is to preview the figure's animated sequence. The program can quickly create previews by reducing the visual quality. For instance, a quick preview of a character walking can be created in a few minutes by reducing the visual quality to black and white and the size of the screen to 360 by 240. You can also check the quality of the fully resolved animation in all its detail, color, and shadowing by fully rendering single frames, chosen from representative parts of the animation, and checking them in detail.

At some point during the testing process and before the final rendering, it is necessary to get approval from the client. The completion of all objects, characters, scenes, and motion is a good time to bring your client to your studio to approve the work. Such an approval can be considered a **milestone** in the contractual process. In fact, most animation companies make this point in the process the specific contractual time to request the second of three payments for the animation.

Rendering

Once you have tested an animation's elements and corrected all apparent mistakes, you are ready to render. Rendering is the creation of each individual frame, complete with all the objects, props, lights, special effects, and motions. The animator initiates the rendering process by setting an array of controls that establish values for certain parameters. These include the resolution and dimensions of each frame, the names and storage location of the frame sequence, and, if the operation is taking place on a network, the computers on which the sequence will be rendered.

Once the rendering starts, the process is automatic and rarely needs operator attention. Large animation studios network dozens of computers together into **render farms**, which can crank out complex animations in a fraction of the time it might take a small desktop studio to render the same sequence.

Obviously, the process of managing your work flow, from the design stage to the rendering stage, can have a lot to do with your studio's efficiency. If you do not plan carefully, you will end up mismanaging your assets and wasting time. For instance, a wise animator will plan a lunch break to occur just when an hour's worth of test rendering is required so that he doesn't have to sit and do nothing while this task is completed. The real trick in a studio with limited computing access is to end the day just when the big, overnight renders are ready to be started.

Often it is necessary to execute several renderings in order to achieve a perfect animation. If a long rendering of, say, 800 frames reveals a mistake between frames 300 and 400, you can rerender just those frames and overwrite the original sequence of frames. You can also use an editing system to assemble the correctly rendered pieces of animation into a final video master.

Getting Paid

Although the overwhelming majority of clients will pay their bills on time and thank you warmly for your efforts on their behalf, you will find a few clients can cause you extreme duress by delaying your payment or not paying you at all.

Your collection job at the end of your animation work is to efficiently execute the closing of your contract by receiving every penny owed. Although a contract for production, entailing the fee and the various penalties for not paying it, is invaluable, unpaid fees due on an animation may not be sufficient to warrant the cost of legal action.

Your best tools of defense therefore are **copyright protection** and **possession**. If your contract stipulates that you own the copyright to the animation, at least until such time as all fees are paid, you are granted extraordinary rights of claim by the U.S. government, namely **treble damages** against anyone using your work without owning the copyright. This increases the amount of fees recoverable and may make legal proceedings worthwhile, though not guaranteed or speedy.

Keep in mind that in addition to the video of the actual animation, other products may or may not be deliverable to the client, as defined by your contract. These include the rights to any characters you create, the source code of the animation, particularly any code you authored from scratch, and the soundtrack, which may include copyrighted music and the artistic work of voice-over talent.

Possession is, by all means, the most valuable form of protection from fee beaters and deadbeats. Simply put, you hold all the useable versions of the animation until all bills are paid. Give the client already half-resolution **Audio Video Interleave** (**AVI**) files (unless the end product is a half-resolution AVI), VHS copies with a **burn-in timecode**,[1] or still photos of the work. When it comes time for delivering the BetaSP version, however, you should hold your hand out and say, "Pay up first."

WARNING

It is important to distinguish between your work as an artist, in which the law assumes you own the animation and all work products associated with the animation, and a work for hire, in which the client owns the animation, the rights to all derivative work products and those contained within it. If you are an employee of an animation studio, your work is most likely defined as work for hire. If you are an independent

---

[1]Burn-in time code: This is a visual representation on screen of the running time of the animation in frames, seconds, minutes and hours, usually shown as four pairs of numbers, separated by colons (00: 00: 00: 00). It is useful in test renders to identify every frame of an animation, but makes the animation nearly useless as a televised element or program, because the numbers are so distracting.

*animator, the law assumes you are the work's owner, unless the specific phrase, **work for hire** is used contractually to define your work. Be careful if this phrase appears in your agreement, because it severely limits your ability to use the ownership of the animation to collect your fees.*

Getting Your First Job in Animation

We've finished the introductory portion of this book and introduced the first technical topic, the workflow process. Now it's time to get you into an animation studio so you can start learning on the job.

"What?" I hear you shouting incredulously, "Me? Get a job as an animator? I don't know a key light from a key frame. What could I do?" That's just the question a potential employer would ask you, isn't it? "What can you do?"

Even if you know absolutely nothing about animation (and chances are you have some knowledge), you can still answer phones, make coffee, sweep floors, write letters, and make calls to collect on delinquent invoices. You can also be a valuable asset if you have some related skill, such as being able to tune a network, defrag a hard drive, or repair a computer. Other important skills include the ability to build a render farm, design a web site, write proposals, make sales (did I mention salespeople are always in demand?), or work on any of the ancillary software products that animators utilize, such as Digital Fusion, Photoshop, or Sound Forge. Or you can offer some of the traditional animation skills such as sculpting, illustrating, modeling, plotting, storyboarding, coloring, or scenic designing.

Any one or more of these skills answers the question, "What can you do?" But before you answer the question, someone has to ask it, someone who can open a door for you. If you don't already have a job in the animation market, your job is to get one. Here's how to do that.

The Issue of Schooling

Let me be really honest here in a way that professors, friends, and especially parents may not be. Animation is an *artistic* trade. Although one could easily spend four years of college or university time and in the process learn to be an animator, animation itself is not a professional academic vocation. Some would vigorously argue with me on this point. However, it is possible to learn nearly everything you need to know about the current animation technology, enough to get your first job and survive long enough to hold it, in six months to a year of full-time study. If academic studies are not your nature, you need not endure them to become an animator. Becoming a manager or studio director, however, may require a degree or two.

You could certainly spend another three years earning an undergraduate degree if you toss in art history, foundation courses in anatomy, sketching, light and shadow, plastic arts, oil and acrylics, and the departmental requisites of a liberal arts curriculum. Judging from the level of maturity I see from most high school graduates I meet, four years could also be spent learning how to wake up on time, make meals, wash clothes, and groom yourself.

If you still need to grow up by spending four more years in school or, more importantly, if you want a well-rounded education with special emphasis on the arts, then a four-year liberal arts education, perhaps followed by a masters, and then perhaps a doctorate, would be necessary. After such a journey, you would be qualified in many areas of professional skills, but you do not need such comprehensive education to be a successful animator.

Strictly speaking, if you really want to be a top-flight animator and still attend a structured institute of learning, you could choose a path that doesn't require a lot of time and money. You always have other options besides shelling out the $120,000 tuition to obtain an undergraduate degree at the **Pratt Institute, the *Rhode Island School of Design* (RISD)**, or the **Parsons School of Design**.

If you are already a mature, focused individual, you might recognize that animation is a trade with a certain curriculum of necessary technical information. If you also have some degree of talent or at least a passion for the craft, you might decide to hone your skills in a trade school, where only your core desire—animation—is addressed. Where might you go to learn 3-D animation, and only that, in such a way as to be maximally trained in a minimum amount of time? The following sections cover four good choices.

Gnomon (www.gnomon3d.com)   Located in Hollywood, California, the epicenter of West Coast digital content creation, Gnomon offers a variety of short and long courses on several areas of entertainment arts, including audio and film production. The advisory board of Gnomon, as well as the faculty, is composed of seasoned professionals who not only lend their considerable expertise to the students, but also serve as mentors and, for those who work hard and apply themselves, initial employers.

Full Sail (www.fullsail.com)   I learned about Full Sail when a young graduate walked into my shop looking for his first job and showed me a one-minute sequence that looked as if he'd been animating for two years. When

I learned that this guy had only just graduated from a six-month diploma course at a place called Full Sail in Orlando, Florida, I called the school. What I learned from this young man and from the phone calls convinced me to add this section to this chapter.

Full Sail offers courses in animation, digital media, film, game design and development, recording arts, and show production and touring. The Associate of Science degree in computer animation includes courses in character design, compositing and editing, shading and lighting. The degree program also offers courses in entertainment and business law, as well as, the most important of all, the business of living/general education, which focuses on the skills needed to get and keep a job in the real world. A degree program takes about 12 to 16 months to complete, based on the program selected. The student body is about 3,000, housed off campus, with a monthly entrance and graduation schedule that avoids releasing large groups of graduates in June and December, thus enhancing the probability of professional placement.

Ringling School of Art and Design (www.rsad.edu)   Located in Sarasota, Florida, the *Ringling School of Art and Design* (RSAD) is a private four-year college that enjoys a reputation dating back to its founding in 1931 as an art institute with a dedicated staff of world-class professionals. Course programs for earning a *Bachelor of Fine Arts* (BFA) include computer animation, fine arts, graphic and interactive communication, interior design, illustration, photography, and digital imaging.

The college maintains a policy of upgrading all computer animation software yearly and all hardware every three years, assuring that its labs will not fall behind the rapidly developing technology. The student-to-computer ratio is 1:2. The student body is about 2,000, half of which are housed on campus.

Future Media Concepts (www.fmctraining.com)   Headquartered in midtown New York, with branch schools in Boston, Philadelphia, Washington D.C., Orlando, and Miami, *Future Media Concepts* (FMC) was the brainchild of a CBS editor and second-generation New York commercial filmmaker. Originally, FMC began as the only factory-authorized training center for Avid nonlinear editing systems on the East Coast, but as demand grew, FMC expanded to include factory-authorized courses in 3ds max, LightWave, and Softimage|XSI.

Courses include nonlinear editing, sound editing, web design, web programming, streaming video, DVD authoring, graphics, compositing, 3D animation, and color correction. Class sizes are limited to a maximum of six students. Weekday and weekend courses are offered along with courses designed for both beginners and advanced-level students.

No set curriculum exists at FMC. Students may enroll in one or more courses the instructor feels the student is qualified for by virtue of previous coursework or a show reel. No degree or diploma is awarded. Prices for courses range from 1-day courses for $895 to 12-day master classes costing over $5,000. The benefits of an FMC course are simply what you learn and the products produced therein, which may substantially improve one's show reel.

FMC is a factory-authorized training center for Adobe, Apple, Avid, Boris, Digidesign, Discreet, Macromedia, NewTek, Pinnacle, Quark, Softimage, Sonic, and Sony. Thus, instructors are trained and qualified by the manufacturers of the software they use. In addition to professional experience, factory authorization ensures that training is sufficient to improve your chances of getting a job.

FMC also offers interested parties an excellent opportunity to learn Avid non-linear video editing software, web design, or digital video by working as interns. FMC interns enjoy free access to all levels of training courses and unlimited practice time on all systems. Internships at FMC are nonpaid positions lasting for three months. Day-time responsibilities include office and administrative work.

No Catch-22—Yet

Animators are luckier than most other media artists. Camerapersons, editors, and directors have to overcome a kind of Catch-22: "You can't get a job unless you have experience, but you can't get experience without a job."

Animators rarely report the existence of a Catch-22 barrier at the doors of their studios. Perhaps it is because computer animation is such a new field and it is growing, both in demand and in technological complexity. This growth keeps current animators from fearing that their jobs are in danger or that they need to closely guard the door of entry to the craft. You can go to a trade convention like *Special Interest Group – Graphics* **SIGGRAPH**, an industry organization, (www.siggraph.org) and easily strike up conversations with the world's foremost animators, sharing job leads and technical tips as if you were chatting with the mailman. You can call an animation studio and ask to speak to the person who hires new animators and chances are you will get the person on the phone. Take advantage of this opportunity.

This is not a business where, if you have no experience, you can't get experience. The trick is to package yourself as a person an animation studio will want to have around and then, while you are around, you gather the experience you need to improve your position.

The Intern

But what if you are a rank beginner with no show reel and no experience? How do you get started? Should you spend thousands of dollars on 3-D animation training courses? Should you major in animation arts in college? Should you just buy the computer and software and work with the tutorials that come with the programs?

Yes.

All these methods are viable ways to become a proficient animator. You can try any or all of them, but a simpler way is available that costs you nothing but your time, and if you apply yourself, this method will make you an animator faster than any other. In fact, if you already have some animation experience—you've taken the $5,000 course or you've been through the college thing and already have a show reel full of tutorial stuff from the manufacturer's tutorial book—this method will work even better.

It's called **interning**. Artists have been using this method to gain valuable experience and obtain true expertise for hundreds of years, since the days when masons built Europe's great cathedrals. In those days, the intern was called an **apprentice**. Medieval apprentices would work for a master for several years and then be elevated to **journeyman.** After several more years as a journeyman, an artist was entitled to become a master and start his own studio with his own apprentices and journeymen.

But you don't have to wait years to get to be a professional animator. Chances are, if you work hard, study, and be a bit aggressive, you can go from intern to paid professional in just a few months. After a year or two, you can open your own animation shop and be master—of at least your own fate.

Being an intern means you make little or no pay, but you show up at the jobsite and work just as if you were paid—in fact, a bit harder, because in addition to working the regular job, you are also training yourself.

So let's start off this book by getting you an intern position.

Six Steps to Getting Your Foot in the Door

Here are the six simple steps you need to follow to get your foot in the door:

1. Make a list of studios you could call.
2. Call each studio on the phone or email the studio manager regularly (but not annoyingly) and offer your services as a free intern.
3. Attend a job interview.

4. Once you start working, be reliable and work hard.

5. Be alert to opportunities.

6. Once you've established your value to the workplace, ask for a letter of recommendation.

Okay, let's examine each step in detail.

Step 1: Make a List of the Studios You Could Call   To make a useful list of contacts and to mine it for the gold contained therein, you're going to need three things. These are a legal pad to keep minute-by-minute notes, a loose-leaf binder of contact sheets (I'll show you how to make them in a minute), and a daily appointment book like a Day Timer (see www. daytimer.com). Get the pad and appointment book right now, then come back, and we'll make the contact sheet together.

Contact Management   The best way to build and maintain an effective list of contacts is to use a computer program called Contact Software. The two biggest brands are **ACT!** (www.act.com) and **Gold Mine**. If you have a few hundred bucks and a computer, get one of these programs and learn how to use it. They enable you to enter contacts as individual files in a database and to place these contacts on a calendar that reminds you when to call each person. The programs even include a system for writing boilerplate letters and emails, so with a simple click of the mouse, you can make the call, log the information, send a resume, and log a date for calling the person to follow up. On the date you selected, the program will beep and remind you to make the call. Contact software is by far the best way to build a client list and make money selling your skills. If you can afford it, buy it now and skip the next part of this chapter that tells you about contact sheets.

I'm assuming, however, that you're starting out with little money. If you don't have the scratch to buy contact software (or the computer to run it), you can do a pretty good job of building and maintaining contacts with a simple method I used for 10 years. I built my business from nothing up to about a quarter of a million dollars a year in billings.

The Contact Sheet   The contact sheet is simply a standard sheet of copier paper, with some fill-in-the-blank spaces. With three punched holes, it could be put into a loose-leaf notebook.

Here's how a contact sheet can look:

Studio Name:

Street Address:

City, State, Zip:

Phone Number

Specialty: Web site:

Contact A: Phone:

Title: Fax:

Contact B: Phone:

Title: Fax:

Contact C: Phone:

Title: Fax:

Contact Notes:

Notice that the contact sheet begins with the studio's name, address, main phone number, and specialty. Next come spaces for three individuals who work at the same address. Don't put any more than three individuals on a page. Below the company data and individual data put the contact notes. This area should take up about two-thirds of the page. It is composed simply of horizontal lines, such as notepad paper. The contact notes area allows you to make notes on the three individuals listed above. Over the years, you are going to fill in a lot of information on these lines. If you learn about other individuals at the company, add a new sheet with their names.

When you make contact notes, start a new line with a letter that stands for the name of the person. For instance, if you first make contact with the B contact, start the first line with B: (don't waste space writing the person's name) and then keep that line just for B notes. Each note should have the date of contact and what happened during the call. Most of the time, you leave messages. If so, write, "B: 12/2/03, LMCB" for "left message to call back."

If you actually talk to B and she says, "Send me your resumé and call me back in two weeks," you have to be careful. Write, "B: 12/15/03, spoke, asked for res, and CB in 2 weeks." Now you have to go directly to your date book (you have a date book, right?), count forward two weeks, and make a note to call B at XYZ Company and follow up on resumé. Next, you have to sit down and send that resumé. If you don't do these two important things, your hard-earned contact with B will have been wasted. Worse, B could get a bad impression of you if she remembers that you never sent a resume. Being professional in the *process of getting work* is every bit as important as being professional *at* work, or even more important.

By carefully using your client lead directory and your daily appointment book, you can keep in touch with thousands of client leads. This is an inexpensive system for keeping track of your cold calls, warm calls, resumé and show reel mailings, appointments, bids, and sales. It takes a bit of discipline, but take my word for it, it works.

Never throw these pages away. As long as you work, these pages will be a valuable record of your past efforts and you can mine this data for valuable leads as long as the individuals on the pages are still alive (and even after a contact retires he or she can still give you references).

Getting on the Phone   Okay, so how do you work the contact sheets and date book? First day: Start out with 10 of these sheets. You can use your local Yellow Pages or something as extensive as the Internet (available free at your local library, right?). Start off by searching for the kinds of studios that do the work you want to do.

The best bet is to use an Internet search engine and type "animation studio" in the search box. If you're not particular about geography and you're willing to relocate anywhere, don't type anything else. Your results page will contain names of studios from every corner of the planet. For that matter, you might want to look up the words for "animation studio" in other languages you know and type them in too.

Let's say you live in Tampa, Florida, and are more particular about your location. Add the word Tampa or Gold Coast to your search. This will limit your results, but you only need one studio to respond positively to your call. The object is to get at least 10 potential listings. Even in the most remote locations, cable stations, corporations, and small production companies ought to total over 10 prospects. If not, face it: You're living in the wrong place to start work as an animator. This doesn't mean that when you're successful and have a big following of clients, you can't move back to that bubbling brook in Long Eddy, New York. As long as you can connect to the Internet, you can get bids and send in work, but when you're getting started, it helps to be living in a market full of prospective employers.

Step 2: Call Each Studio on the Phone or Email the Studio Manager Regularly (but Not Annoyingly) and Offer Your Services as a Free

Intern Once you have the basic contact information, call the studio's main number. Your objective is to find out who can authorize you to come in and work as an intern. To find out this valuable information, you have to become friendly with the person who answers the phone.

Do you need a script for this? I've prepared two scripts for the person who answers the phone. The first is for small studios of 3 to 10 people, and the second is for large outfits like Industrial Light & Magic. Here's the small-company receptionist script.

> Hi, my name is _____ (your first name only) and, uh, I wonder if you can help me out. You see, I want to offer my services to your studio for free in return for learning 3-D animation programs. Can you tell me the name and title of a person who might be able to use me?

I added an "uh" in there to break up your flow and keep you from sounding like you're reading a script. You can add any form of nervous hesitancy you like, even if you're a professional public speaker, because nobody wants to help someone who sounds so well put together that they need no help from anyone. After a while, you'll learn how to do this without a script and eventually you'll be so confident, you'll be able to vary the script to fit the situation.

If you're calling a large studio (more than 10 employees), use the big-studio receptionist script:

> Hello. I wonder if you can connect me to the person in charge of entry-level creative staff and before you connect me, can you give me the person's name and title please?

Notice that in each script you've asked for the name and title of the person to whom the receptionist is sending you. This is because receptionists are often poorly trained and just connect you to someone without telling you who that person is or what they do. You might be connected to the accountant's phone line for all you know, and if you get his voicemail (as is most likely), you could waste weeks leaving messages for a person who has nothing to do with what you need.

Asking the receptionist for the name and title also forces the receptionist to do a little homework rather than shuttling your call to any bozo whose line is open. Be sure not to be a bozo yourself. Write the information down on your legal pad. That way, when the person answers, you can say, "Hi, Linda. I understand you're the director of the animation department."

Now you give Linda the main-contact script, offering her your time for free in return for the opportunity to learn on the job.

> Hi, _____. My name is _____ (first and last name this time) and I'm calling to offer you my services for free in return for the opportunity of improving my resume. Do you have room in your studio for a hard-working gopher?

Gopher or gofer is a buzzword in the industry. It means the person who'll "go for" this and "go for" that. Most of the time, such an offer will be received with a positive response. The response you want is to be invited into the studio for a face-to-face interview. You can actually suggest this if the call goes from the introductory main-contact script to a friendly chat. Another positive response would be for the main contact to ask you for a resume in the form of a mailed, faxed, or emailed letter. This is a formality. It doesn't mean the person is offering you an internship and it doesn't mean you're not wanted. It's just that the person wants to check you out before taking the next step.

In either case—the appointment or the resume request—you're going to need a resume and a nice cover letter. Let's sketch out each here.

The Resume   I've seen a couple thousand resumes in my time and what's really amazing is the immense range they represent in quality and inventiveness. I've seen a resume printed in four colors on custom paper stock that had six misspellings (not hired) and a resume typed on a manual typewriter on plain copy paper that described a stunning career in two paragraphs (hired).

You can buy a book and study for a month on how to write a resume, but here's a simple formula that has always worked for anyone who graduated from my classes or internships. Just follow the format (not the exact words) and keep it to one page. Nobody wants to get, read, or store multipage resumes. Even seasoned pros, who have created feature films and have 10 years' experience craft, their resumes onto one page. If they can, so can you.

In fact, the trouble is that at this stage of your career you may be stretching it to fill half a page. Don't worry. Nobody expects you to have *Monsters, Inc.* or *The Matrix* on your resume if you're looking for work as an intern. They just want to see that you know the form, can present a neat image of yourself, and have the discipline to compose a page of text and mail it on time. For now, the resume is a formality. Later, it may be a lot more, so get in practice (see Table 1-1).

Table 1-1

The typical
resume format

First Name, Second Name
Street Address
City, State, Zip
Phone: (note if you have a message machine}
Cell: (good idea if you can afford it)
Email

Objective:

To obtain an unpaid internship at an animation studio in return for the opportunity to teach myself computer animation skills on the studio's computers.

*You can revise this as it suits the recipient. For instance, if you are sending it
to a corporate TV studio, change "animation studio" to " corporate production
facility."*

Employment Experience:
Name of Company Last Worked, Address, City, State, Date Hired,
Date Left (or "Present"}

*This is one short sentence to say what you did at this job. It doesn't have to be
animation related. It just needs to show you worked for somebody at some time.
If you don't have anything to put here, just write that you are requesting this
internship in order to obtain your first job experience. After this job, list your previous jobs in reverse-date order in the same format.*

Education:
School Name, Address, City, State, Degree or Diploma, Start Date to
Graduation Date

*If you had some stellar results in school, such as a high GPA or an honors
degree, write one short sentence here; otherwise, don't write anything. After this
entry, include the other schools you attended.*

*If you attend or attended postgraduate school, you should first list your grad-
uate school experience and then your undergraduate college, but not high school.
If you attend or attended undergraduate college, list that first and then high
school, but only if your page is not filled. Otherwise, do not list high school. If you
attended high school and have been accepted to college, list the college first, and*

under the first line write, "Accepted to matriculate in (month/year)." If you attend high school and have not been accepted to or plan to attend college, just list the high school. Many animators start right from high school without going to college. It's a struggle, but it happens.

Credits:

This is a tough section to fill in if you are just starting out. If you've done a few projects for school or cable TV, for instance, list no fewer than three and no more than five. If you have more than five legitimate credits, you should be looking for a paid job, not an internship. Be sure to have these credits on your show reel if you can. If you have no credits, do not include this section.

Personal:

Include two to three brief sentences here to describe such things as volunteering, hobbies, and job-related skills. This is a place to show how your personal life makes you a good job candidate. For instance, if you volunteer to feed homeless people on Sundays, write

Volunteer work: City homeless shelter cafeteria once a week.

If you're a SCUBA diver and like to watch football on Monday nights, write

Hobbies: SCUBA diving and televised sports.

If you speak two languages, know bookkeeping, and have a truck license (valuable skills for any business), write

Job-Related Skills: Fluent in French and Spanish, knowledge of bookkeeping, truck driving license (standard shift).

And if you already know some of the software that is used in our industry, be sure to write
Photoshop, Flash, Microsoft Office.
References and show reel available on request.

Always end the resume with this line (adding "show reel" only if you have a show reel) and have three people other than relatives who will answer a phone call from a potential employer and attest that you aren't a child molester or drug addict.

The Cover Letter   The cover letter is a really simple piece of creative writing. Make it as short as possible, and configure the text to fit dead center in the page. You can use personal stationery if you want, with your name and a clever logo at the top of the page, but do not under any circumstances send any printed matter (a resume, letter, or business card) that implies you own a company.

WARNING

I see a lot of guys and gals that want to go out into the business world presenting themselves as some form of corporate entity. Joe Blow will get out of college and immediately print up business cards with "Joe Blow Productions" at the top. That's great. Obviously, your end goal is to run your own company, but don't go looking for a job at somebody else's company using stationery from your company. Nobody wants to hire a competitor. You'd be cutting your own throat.

Try Table 1-2 for a cover letter:

Table 1-2

Typical cover letter for an entry level resume

First Name, Second Name
Street Address
City, State, Zip

Mr./Ms. Firstname Lastname
333 Street Name
City, State Zip

Dear Firstname:

I would like to serve you as a free intern in return for the opportunity to sit at one of your workstations and teach myself 3-D animation. I am available from Month, Day to Month, Day, and can work from 9:00 AM to 2:00 PM, Monday through Thursday.

My enclosed resume briefly outlines my background, which, as you can see, needs deepening.

Please help me go deeper!

Sincerely,
My Signature
My Name

Following Up   Remember that we started with 10 contact sheets? Whenever you get a genuine contact's name, that's the time to fill out the contact sheet. A genuine contact is a person that is most probably the person who can hire you. This can be a person you identified from a web site, someone from a business card, your dad's school friend, or the person the receptionist named but who wasn't in the office and you left a voicemail message for her. Whoever.

As soon as you have a name, title, and phone extension, fill in the contact sheet. If you leave a message, put LMCB on the Action line. If you speak to a person, write "Spk" and put the person's initials next. Then write some notes to remind you what the conversation was about.

Next, immediately go to your daily appointment book and add an appropriate "to do" on the page. An LMCB might require a call back again in two weeks. A "call me back in the fall" would require a "to do" some months into the future. An appointment, of course, must not be forgotten, so use your appointment book. Always, always, always put a "to do" message in the appointment book at some future date to remind you to keep in touch with the person. Too often a good contact is forgotten because no "to do" was entered. Nothing impresses an employer more than when you remember to call. It shows real professionalism.

Step 3: Attend a Job Interview

If you were diligent and called at least 10 people each day for a week or so, you will have made at least one appointment to see someone about being an intern. Now that you have someone who is willing to see you, set the foundation for a great appointment.

Don't go empty handed. Prepare something to hand to the person when you walk in. This can be another copy of your resume, which is a good idea, because most people will lose it (or write something on it they won't want you to see, such as "see this person when you can").

If you have some new piece of work, a video, or even a sketchbook, bring it to the meeting. It may not be a masterpiece, but it could start the conversation flowing and make the interviewer more comfortable with you.

If you are going to bring videotape or projection materials, be sure to ask your host in advance so that the appropriate equipment can be set up for your presentation. If it is more convenient to make your presentation in a conference room, be sure to ask your host to book one for the meeting.

Finally, if your appointment is more than a week in advance, be sure to call the client the day before the meeting to confirm the appointment. Often a client will simply forget about the appointment or develop a conflict without informing you. You don't want to dress up, give up part of your workday, drive an hour, and find out the client had to go to Kabul at the last minute but didn't call you to postpone the meeting.

Be on Time   Most people do their best to arrive for a job interview on time. Arriving late is one of the worst things you can do. If you're going to be late for a meeting, at least call ahead well before the appointment time. Bad things happen, but you have no excuse for being impolite.

Dress Right   I know what you're thinking. The old guy is going to say, "Wear a suit." Nope. Wear whatever makes you comfortable. Being comfortable in a job interview is important. Keep in mind, however, that the person interviewing you will be dressed for his or her comfort. If the person looks a lot different than you, it will make you uncomfortable. A suit, for an animator, is definitely bizarre. Animators wear ripped jeans, wampum belts, nose rings, Mohawks, that kind of stuff. And these are animators who *already* work at the studio. They paid their dues. They are known talents. You are unknown. If you walk in with a bad dress code attitude, you may make someone uncomfortable and that will work against you.

Go safely dressed. Wear plain slacks, a dress shirt, dark leather shoes, and no piercings or obvious tattoos. Consider yourself a mystery that will be revealed to your employer later when you are admired and respected. For the interview, dress safely.

Step 4: Once You Start Work as an Intern, Work Your Tail Off

Getting an intern opportunity gives you the chance to accomplish two important tasks: getting experience and getting hired by the firm for which you are interning. How can you accomplish both these tasks?

Here are five good tips for a successful internship:

- *Work your tail off.* The more work you do, the more you are going to learn. From your first day on the job, ask what you can do to be helpful.

Let's say your supervisor gives you a job to do and then leaves you alone to do it. When you're finished with the job, don't sit around doing nothing or listening to your headset. Ask for another job. If you find yourself without work, look for something helpful to do. No place is ever clean enough. Collect the empty coffee cups and donut bags. Be useful.

■ *Be pleasant.* Everyone will know you're an intern and you're not expected to know much or be a great conversationalist, but be nice. If you're going out for a break, ask if you can run an errand for anybody. Don't gossip or talk a lot. Listen to everyone around you and get a feel for how people in the shop behave. Blend in.

■ *Seek opportunities to learn more.* Many media production shops have idle computers with lots of interesting software on them. Ask if you can use a computer to teach yourself some software. Get the documentation (or buy a tutorial cheater book) and learn a new program.

Interns at my company get little guidance; I haven't the time. I just give the intern one computer and tell him or her to use it to learn anything he or she needs. I have a library of documentation and tutorials, and I expect that idle interns will use these tools to expand their knowledge and become more useful. Occasionally, I'll walk through the studio and look over an intern's shoulder. If I see something interesting, I'll lend advice. Maybe I'll see something I can sell a client. If I do, the intern is considered hired as a freelancer on that job. If the intern just sits around and does nothing, no such opportunity would arise.

■ *Learn the rules of the shop and follow them.* Can you use the phone for local calls in your spare time? If not, don't. What time are you expected to show up for work? Don't be late. Interns are not usually paid employees, but if you say you're going to be at work on a certain day at a certain time, someone may be relying on you. If you can't make it, or you're going to be late, call ahead. Pretend this is a real job. Are you living up to the expectations of a real job?

■ *Find a mentor.* A mentor is someone in the shop who will take an interest in you and help you out. Every good workplace has at least one person who likes to help out young, eager learners like you. Find out who this person or people are, and do whatever you can to help them out. Often, the older people in a shop are great mentors, but sometimes you can find a person your own age who has a little more experience than you and who is not averse to showing you around.

Finding a good mentor can boost your career significantly. Experienced people enjoy being mentors for lots of reasons, but the main one is that they believe what goes around comes around. In other words, when you get some experience yourself, try to return your mentor's generosity.

Step 5: Be Alert to Opportunities

While you are spending time as an unpaid intern in a busy animation studio, lots of action will be taking place around you. Clients will be coming in, proposals will be written, projects will be started, storyboards will be created, scenes will be developed, and frames will be rendered. Throughout it all, opportunities will occasionally present themselves like little angels on your shoulder.

Let's say the assistant animator decides to take a lengthy sabbatical in Peru. Opportunity. The computer network software you just learned has a glitch. Opportunity. Your boss is ready to turn down a job because it doesn't pay enough. Opportunity.

Opportunity for what? For you to jump in and save the day, that's what.

The better prepared you are for professional-level work, the better able you will be to take advantage of opportunities. For instance, the assistant animator suddenly leaves. Will it be on the day you just completed studying the software he was using, or will it be on the day you finally got a perfect score on an online game?

Of course, not every opportunity will fit your abilities. If you think it does, speak up and say so, but only once. If you don't get a positive response, forget about it. Maybe your boss already had somebody in mind for the job before you arrived. Don't show any disappointment if an offer to exploit an opportunity doesn't go your way. Others will occur. The fact is the opportunities that usually happen outnumber the qualified people who can exploit them.

Step 6: Once You've Established Your Value to the Workplace, Ask for a Letter of Recommendation

It should take you at least a month of full-time work to establish yourself as a known commodity at your workplace. If you are good at your work and extremely fortunate, management may even offer to pay you. You might be hired as a salaried employee or be offered some form of **freelance** work. Although the definitions of **salaried employee** and freelance are legally determined on a state-by-state basis, the general understanding is that when an employer wants absolute control over you, he or she will offer to make you an employee and pay you compensation consisting of salary and **fringe benefits.** These benefits could include healthcare, pension, stock options, low-interest loans, death benefits, a company car, and so on.

When an employer is principally concerned with getting a specific job done and is not concerned exactly how the job is accomplished, he or she may offer you work as an **independent contractor** or **freelancer.** Legally speaking, the difference between salaried employee and freelancer is determined by the amount of control the employer has over you. In some states, even one day of paid work can be legally interpreted as a salaried relationship, wherein the employer is obligated to provide you with a complex formula of benefits, such as unemployment insurance and workmen's compensation. You may want to become familiar with the laws in your state and can easily do so by contacting the appropriate state offices or visiting their web sites.

The important point to remember, however, is that when you are offered employment, a significant change occurs in your relationship to the workplace. Employers are not likely to make this change lightly, and rarely will they do so after only a month of full-time internship. Therefore, do not be too concerned if you do not get offered the Holy Grail of employment after you have dedicated yourself so many hours for free.

So what *can* you expect after a month of full-time work? By this time, you should have learned at least one piece of software, done some kind of service project for your company, and become familiar with your coworkers. More importantly, management should have come to know you as a reliable commodity. Now is the time to plot out your immediate future. Choose a time when your manager is not too busy and ask for a few minutes' time. If you're in a really good shop, your manager may have even set up a meeting like this without you asking. Here's what you say:

"I've been here now a month or so, and I wanted to know how you think I'm doing."

This opens the door for the manager to give you praise or criticism. Listen to both carefully. For the praise, respond with a simple thank you. If criticism is made, pay close attention, and again thank the manager for giving you this valuable information and a chance to improve. For each criticism, do three things:

■ In your own words, repeat the criticism back to the manager, so he or she knows you understand. Also ask for confirmation, such as, "In other words, you're telling me that when I'm going to be late for work, I should call you? Is that right?"

- Sincerely apologize for the error that led to the criticism, such as, "I'm sorry I was late two times this week and didn't call when I couldn't come in yesterday."

- Assure the manager that you won't allow this to happen again, and explain exactly what you are going to do to prevent it from happening again. "I'm going to buy myself a new alarm clock tonight and put your phone number on my cell phone directory."

You have now brought out, between you and your manager, the good and bad points of your work. In other words, you have had your first "performance evaluation." This is a good first step. You are now practicing professional behavior.

If your criticisms were significant, take another month of interning and have another meeting with your manager. Keep up this cycle of education until the criticisms are gone or have diminished to zero or really minor issues.

Remember, the object of becoming an intern is to learn the craft, get job experience, and eventually land a job. As an intern, you are getting the first two objects from day one. But what about getting a job? Should you come right out and ask for a job? No.

The very fact that you are performing duties in a professional workplace voluntarily, without pay, is already a powerful statement. In every moment on the job, you are silently declaring, "See what I'm worth? Will you hire me?" If after a reasonable amount of time, during which you have been doing satisfactory work, the employer isn't responding with an offer of compensation, the employer is just as clearly answering you. He is silently declaring, "Nope." So why bother asking for work out loud? Do you want to hear the nope spoken?

Although most interns start to work at a production company with the expectation that they will one day get put on the payroll or at least be paid as a freelancer, this simply may not happen. The shop may not be busy enough to afford you. The manager may have a nephew graduating from design school next month. As hard as it might be to accept, your work may just not be good enough yet. Sadly, interns are often looked upon as "just kids." Even though you may be professional and ready for a paid assignment, it is possible you could be unfairly stigmatized by the shop's professionals simply because you came to them first as an unpaid intern.

It may come to pass that for one reason or another, the shop you are interning is not the right shop for you professionally. No matter how good you are, you will never get hired.

If you should reach the conclusion that this is so, or if at any point you feel that it is high time you receive some pay or leave the place, don't ask for pay before you ask for a letter of recommendation. Simply ask for a private moment of your manager's time and say something like this:

"I'm really happy working here and I want you to know that I am very grateful for the opportunities you have given me to increase my professional experience. I was thinking it's time for me to start looking for a real job while I continue my internship. I was wondering if you would be so kind as to write me a letter of recommendation."

It's as simple as that. In most cases, the manager will be glad to write you a glowing letter on company stationery, which will greatly improve your chances of getting work at another animation shop. If you were a great intern, your manager may even be enthused enough to go beyond the letter and offer you some leads or even make some calls for you. Even if you weren't the best intern, the manager will still be glad to write you a recommendation because it will increase the probability that you will promote yourself out of the shop without the manager having to suffer the pain of telling you to go. In other words, you'd have to be one really bad intern or be working for one really terrible boss to be refused a letter of recommendation.

Now that you have enough business information to start you off on the path to full-time employment, let's open the door to the technical information you need to understand the craft of animation.

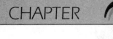

From Concept to Screen: The Workflow Process

This chapter describes the typical sequence of tasks required to complete an animation project, and the chapter's case study is a 60-second commercial. Take a moment to look at the commercial on the CD that came with this book.

Let's pretend you're an animation studio owner or a top-level manager in a production facility where you run one or more animation projects. A project comes to you on day one, and you've got to plan and execute the job within a given amount of time. Your success depends on your ability to manage resources (labor, computers, and supplies), time (from day one until the deadline), and money (the budget for the project). What is the work process you must initiate to achieve success?

Conceptualizing the Animation

Often an animation studio head or a lead animator is required to assist the client in conceptualizing the animation. Sometimes the client may not even know animation is the solution to the problem. The conceptualization process might not call for animation, or it might include the possibility of live-action video, photography, and even live performances. The more experience an animation producer has, the more useful he or she can be during client contacts. Obviously, client contact requires the highest level of both animation and personal communications skills.

Conceptualization is the process by which a creative vendor helps a client achieve a communications goal. If the client wants to sell a designer shoe, for instance, the conceptualization may take the form of a commercial or a website page. In either case, the concept must identify the *unique sale propositions* (USPs) of the shoe (it's fasionable, its soft) and present them in a memorable message. This message might be an animation, live video, graphics or a combination of all three.

As a creative vendor, it's your job to choose the best combination of media, images, statements and styles to achieve the client's goals. If your expertise is animation youu will be able to address a fair percentage of the requirements of a conceptualization, but not all. Therefore you will either subordinate your efforts to a creative director who has broader experience in all media, or you will only meet with clients who require animation concepts.

At such meetings, and the sessions that follow them, your job will be to solve problems (or entertain an audience) with animation. A client may have a toy that is selling poorly west of the Mississippi or a skin cream that employs a new scientific process. Your job as a conceptualizer is to first understand the client's problem and be able to address it with graphic solutions. Once you think you have the solution, you are required to present it effectively to the client, who will then have to decide if your solution is worth the money to execute.

As an animator, your solution will tend to be animation oriented, but you must also keep your mind open for other media that, when combined with animation, will make the solution even more effective. Knowing how to combine animation with video, film, web site content and even print applications will enhance your effectiveness. And lead you to become a creative director. Your role as an animator therefore may be part of a larger production or a stand-alone animation.

Sooner or later, your job as a conceptualizer must include being able to sketch your ideas with a pencil and paper. You may need to draw a cute character and convince your client that the character will win him or her more customers. You may need to sketch a forest setting or an urban street where your animation will be staged. Finally, you will be required to create a full, graphic depiction of the entire animation—a series of drawings, a lot like a comic book. This is called a storyboard.

The Storyboard

The **storyboard** is the blueprint of an animation. Depending on your role in the production, you may be required to create the storyboard from scratch, adapt and modify an existing storyboard, or be given a storyboard that is complete and over while you have little or no control.

I'm going to use a real case study from a recent production.

Conceptualization This project was for JVC's new line of video projection systems called D-ILA. The client expressed an interest in creating a one-minute animation that could introduce the product's slogan, "Images of Perfection." The animation itself would be shown on a D-ILA projector at the **National Association of Broadcasters** (**NAB**) convention in Las Vegas, the largest technical tradeshow in the world. The animation had to be completed in 50 days. Because the projectors were capable of **high-definition** (**HD**) resolution, the animation would have to be created in the **1080i HD format,** requiring interlaced frames measuring 1,920 pixels wide by 1,080 pixels high. Given this challenge, I returned to my studio and had a meeting with **Jack Ehrbar,** then my chief animator.

Because our meeting only involved the two of us, we sat at a table with a large pad of white paper. Conceptualization meetings in larger studios or advertising agencies can include dozens of people, so a whiteboard might be used. The paper or whiteboard acts as a kind of visual kick start, its emptiness prompting you to get thinking.

The first task of conceptualization is to list the objects of the animation. What does the client want out of the project? These objects might start at the basic level:

- Attention to the new product
- Immediate sales leads
- Product brand recognition
- Corporate brand recognition

I usually put basic objects on the board or paper with lots of space beneath, because once written they demand further details, such as the following:

- Attention to the new product
  - USPs
  - Competitive features
  - Weak points
  - **Focus group** results[1]
- Immediate sales leads
  - Geographic-specific issues[2]
  - Direct appeal to action
  - Consistency with client **collateral**[3]
- Product brand recognition
  - Preexisting product characters or elements
  - Preexisting image of the product

---

[1]Clients who want to know what their customers think about a product often conduct focus groups. The client, usually anonymously, assembles a group of customers and has a professional interviewer solicit opinions. Without knowing which company is sponsoring the group, customers often give frank appraisals that reveal weaknesses and strengths about the product.

[2]These are concerns that relate to the location in which the animation will appear. Often, viewer expectations with regard to race, accent, subject matter, and other demographic issues play a part in animation design, particularly character design.

[3]Collateral consists of material the client is distributing in addition to your animation. For instance, if the object of an animation is to get the viewer to order a videotape, should the animation design be adapted to the graphic appearance of the video and its package?

- Radical animation idea
- Humor
- Clever characters
- Corporate brand recognition
  - Corporate image
  - Logos and design limitations
  - Legal issues

Straight away, as an artist, you might look at this outline and think, "What does all this have to do with a creative design?"

A lot. Despite what you might think about society's debt to you or your entitlement to enjoy a life of fulfilling, enjoyable work, the basic fact of life that underscores almost all animation work is that our work is used for practical purposes.

Aside from pure artwork that you might create **on spec** (that is, you don't get paid up front) or under a (rare) grant or commission, your work is going to be contracted for and judged by its commercial value. That's why it is important to think commercially before you create. The closer you can get your creative mind around the specific business goals of your client, the more completely your end product will satisfy your client. In the commercial animation business, total client satisfaction is the main goal.

Learning how to organize your creative thoughts along a predetermined course of necessities is not such a bad habit to acquire. If you don't believe me, consider how many times you've sat in front of a blank screen or paper totally lost for an idea. Try making a personal satisfaction list of what you want out of your animation similar to my client satisfaction list above, and I guarantee the ideas will flow a lot faster. In fact, you may have to discipline yourself to finish the list as ideas start flowing from your first notes.

So Jack and I are sitting with this list of client satisfaction issues for the JVC animation, and we can see some important things forming up right away. Here's how our list developed.

Under Unique sales points, we wrote "Excellent images, advanced design chip." Under Competitive features, we wrote "Best resolution of any projector, no moving mirrors on chip." For Weak points, we wrote "Relatively expensive, bulb lasts 1,000 hours, costs $1,000, $1.00 per hour to run." Under Focus group results, we wrote "None."

For Immediate sales leads—Geographic-specific issues, we wrote "Best to sell in large metro markets, especially New York and LA." Under Direct appeal to action, we wrote "Get everyone at NAB to watch the following video sample reel." For Consistency with client collateral, we noted "JVC

Brochure has a full-page logo we should incorporate as closing frame in animation."

Under, Product brand recognition—Preexisting product characters or elements, we noted "Product Slogan: Images of Perfection. Corporate Slogan: JVC Takes You Further." I also noted, "Let's try to find out who actually owns the Nipper the Dog image?" Jack added, "Egyptian themes popular in current cinema releases (*Mummy II*, cover of *National Geographic*, Indiana Jones)."

Under, Preexisting image of product and Radical animation idea, we wrote "No controls here. Let's see how far we can take the client. Humor has to be tasteful and bicultural (JVC is a Japanese company). Clever characters, but only if they support the basic sales theme."

Under Corporate brand recognition—Corporate image, we noted "Excellent engineering at an affordable price. Always bringing out new technology, but not enough money behind marketing. Need new image for acquired projector line." (JVC acquired a division of Hughes that brought video projectors into JVC's product line.)

Drawing the Storyboard   With all of this now on paper, Jack and I let our creative juices flow. Each idea was filtered through the statements we had written in our list of objectives. I was keen on using the Nipper the Dog image. This is a really old logo that goes back to the Victor Company and features a little dog staring into the megaphone of a wind-up Victrola record player. Under the picture is a slogan, "His Master's Voice." The idea behind this logo is that the Victrola can record a person's voice so well that the dog will think it's real.

My idea was to bring this logo up-to-date and have a dog looking at the screen of the D-ILA projector and thinking it was really his master. The slogan would be "His Master's Image." I went off to start sketching storyboards that had the dog running repeatedly at the screen, bouncing off and running back. Finally, the dog, tired out, lifts his leg toward a fire hydrant that appears onscreen.

Jack took a different approach. He focused on the Egyptian theme and sketched out a storyboard that was reminiscent of Indiana Jones. In Jack's story, a group of Egyptians led by a Dr. Jones tries to break into a pyramid searching for "Images of Perfection." Finally, in the innermost chamber, the explorers find a living pharaoh watching scenes of his pyramid on the D-ILA projector. This concept especially countered the bulb-life issue.

I'm not as good a storyboard artist as Jack, but after we were finished, we critiqued each other's work, pointing out weaknesses, adding ideas in order to stick to the list of objectives, and otherwise honing the results. Eventually, Jack finished the two boards in his highly professional hand.

The leg-lifting scene was deleted. It wasn't in the best of taste obviously. I also had to call lawyers and the Library of Congress to find out if JVC did or did not own the logo. JVC thought it was still owned by RCA but were not sure.

Jack's idea got revised a bit too. We took out the crowd of explorers and honed it down to one Egyptian and Dr. Jones. The idea worked just as well and it's easier to animate two characters than eight.

Once the storyboards were done, we set an appointment with the client and then started to write dialog for the characters in our boards.

Adding Dialog It's not often a good idea to put dialog in storyboards, unless absolutely necessary for the concept to work, or unless you are going to send the boards to someone without the benefit of a live presentation. This is done for two reasons. First, a really good animation board should work without dialog. This is not to say that dialog is bad. Many great animations can't work without the characters talking. It's just a good test of your visual abilities to try to create boards that need a minimum of dialog.

The second reason is that clients tend to criticize a board based on text. Text is easily picked apart. As far as pictures, most clients can't draw, so they don't tend to be critical about pictures. They either like them or they don't. So if you attach too many words to your pictures, you are giving the client a chance to criticize and perhaps reject your work. With animation boards, it's so much better to go in to a meeting and talk your way through the dialog. That way, if you see a frown or read displeasure, you can switch your direction and change the dialog on the spot, that is, if you're creatively fast on your feet.

My storyboard, admittedly the weaker of the two, had some dialog to show that the image of the owner onscreen was calling his dog. "Here, Nipper!" That was it. After bouncing off the screen a few times, Nipper sits by the projector and pines at the screen. The slogan appears below, "His Master's Image."

Jack's storyboard had the professor saying, "Sabu, we must find our way to the inner tomb to find Images of Perfection." Later, as Sabu presses several secret stone panels, the tomb opens itself and Sabu exclaims, "Aiee! The power!" The professor joins in, "The power to take you further!"

The dialog on Jack's board was more extensive, but we felt that because it supported the client's slogans, there would be no risk to include it.

The Client Presentation

Sooner or later, you are going to have to present a board, hopefully your own, to a client for review. Often, your ideas will be in open competition with other vendors. This is no picnic.

Do your homework and think of every possible variation on your idea, so you can be prepared to switch direction if the client isn't agreeable with what you've done. Even take a pencil and a blank pad in case you need to resketch on the spot.

My presentation with JVC went pretty much along normal lines. The main client was there with the product manager for the projector line and another executive who managed the engineering aspects of the projectors. I opened the presentation by saying that my company had prepared two ideas for review and that any other ideas the executives had would be welcomed.

During the week before the meeting, I thought I had ascertained that the Nipper logo had been abandoned by RCA and that ownership had indeed reverted back to Victor, now known as the Victor Company of Japan, or JVC. I was wrong. The executives, who were aware of my search in advance, had consulted with their legal counsel in Japan and learned that they could not use the logo. I learned this as I presented my first board, the one I had designed.

The executives loved the idea of the dog bouncing off the screen, but they weren't sure whether they could get away with using it, even though they felt it was a fair and great spoof of RCA and their own corporate history. Their response, which was all positive, got me enthusiastic and ready to present Jack's board, which I knew was the stronger concept.

Presenting the Egyptian idea took a bit of time. This board had more details, which lead to several questions. What kind of pyramid can we use? Do we have to worry about Middle Eastern politics? Is it politically correct? At this time, I began to worry. What if legal trashed the doggie idea and they didn't like the Egyptian idea? In the time it would take to make new boards, the job's funding could go south. Another company could get in with a different idea, such as a live video, and maybe the budget would get absorbed by other concerns. This was not good.

Then the head honcho of the tradeshow asked, "Where is the company history?" The other executives stopped, "Company history? We need a company history?"

"Sure," the honcho nodded. "We have a pyramid here. This is history. Remember? We did a print ad some years back with a pyramid."

"Yeah, I remember," said the engineer, smiling, "That was a good ad."

"And it had JVC's history in it. The invention of the first Japanese TV set by Mr. Takayanagi. The invention of VHS. The first camcorder. History." The honcho turned to me, suggesting a deal was in his head. "Can you work our history into this?"

Out came my pencil. Where the Egyptian character was pressing on one of the tomb's secret stone panels, I drew in a picture of a Japanese man's head. Explaining as I sketched, "We do this like hieroglyphics. This is Mr. Takayanagi."

Then I went to another panel where the Egyptian is pressing another stone, and I sketched in the letters "VHS," but kind of like they would appear in hieroglyphics, with the V being a bird and the S as two beetles.

They started to get the idea and started laughing and pounding their hands on the head honcho's back. "The Egyptian guy can be saying here, 'Dr. Jones, they knew about broadcast! Look! Dr. Takayanagi!'"

This split the executives up. A twentieth-century Japanese engineer in a 3,000-year-old pyramid? This was funny. This worked. Then when I showed them the last panel, where the pharaoh is found in the tomb, the executives started to get excited. "Can the two explorers shout something like, 'JVC takes you further?'" one asked.

"How about this?" I suggested. "As they start getting closer, we hear a rumbling building up inside the tomb. The little Egyptian guy starts shaking and saying, 'Aye, Dr. Jones! The power! The Power!' And the professor, not catching on at first, asks, 'What power?' Then he sees the doorway glowing—get this guys—with the D-ILA logo!'"

"Yes, I like that," said the engineer.

"And the professor knows what power the Egyptian guy is talking about," I add. "Just then the Egyptian guy presses the last stone and the door opens, and tons of light, because the D-ILA is a really bright projector, tons of light come out at them and they stagger back, yelling together, 'The power to take you further!' Then we see the pharaoh munching popcorn and he laughs and says, 'D-ILA.' Nothing more."

"Yeah, that's it. That's the way we want it." They all nodded together and we all started working on small details. At this point, of course, I realized that I had sold the board. Because I already knew the budget from a previous meeting I had with the client, I proceeded to carve out the parameters of a contract: how long to deadline (50 days), what format (HD 1080i), and how long the animation would be (60 seconds).

The executives agreed that the second board was what they wanted, and we shook hands, made arrangements to finalize a contract in a few days. I left the meeting and headed back to the studio to start **timing the board**.

Timing the Board

Sooner or later, your board is going to have to get translated from flat, static paper to dynamic moving video or film. Commercially speaking, the best time to add the element of time to a storyboard is after you've sold it or raised the budget to produce.

This is because the element of time brings with it the element of expense. Until now, your expenses have been nearly nil—a pencil, some paper, and lots of sweat. Now, as you add the element of time, you have to start thinking about how many frames will need to be rendered, how many characters and scenes need to be designed, and what kind of computer you will need to execute the job (or, if you're a traditional animator, what kind of camera and what kind of film you will need). In any case, your expenses are going to start piling up.

In addition to producing the visuals, you will now have to consider audio. Will you need actors to perform the voices? How will you synchronize their voices to the characters' lip movements? Will there be sound effects? How about music? Will you write and perform your own, pay a composer and musicians, or purchase stock music clips?

All these issues relate to timing the storyboard. Start out by deciding how long the entire production will be. Figure 2-1 demonstrates the relationship between time and the amount of frames you need to create. The number of frames required for a production depends on the frame rate. If outputting to film, the rate is usually **24 frames per second** (fps), while video is customarily 30 fps.

Once you have settled on the final length of the production, you can start to allocate the amount of frames to specific panels of your storyboard. A 30-second commercial spot, for instance, might be expressed in a storyboard

Figure 2-1
Frame rates

Length of animation	24 fps film rate	30 fps video rate
30 seconds	720	900
60 seconds	1,440	1,800
3 minutes	4,320	5,400
20 minutes	28,800	36,000
90 minutes	129,600	162,000

with 10 panels. On the average, you would have 72 frames to express each panel, but this is only an average. Many boards use several panels to describe a sequence of action that, in the final spot, may occupy only a small percentage of the final length.

For instance, a storyboard for an electric toothbrush might have 20 panels to carefully demonstrate the inner workings of the brush and how it cleans teeth, with only 5 panels to show the woman using the toothbrush and how the box looks on the store shelf. In reality, the 20 technical panels might occupy only 10 seconds of the final spot, whereas the other 5 panels represent the remaining 20 seconds. It would be misleading therefore to divide each panel into 72 frames. This is why you must carefully break down the time of a storyboard and indicate the amount of frames under each panel. In this way, everyone who works on the board (even if "everyone" is just you) will have an accurate starting point of design (see Figure 2-2).

Once you have assigned specific times to each panel, you may realize that some panels will have to be expanded, that is, more details will have to be supplied, before the actual animator, if it isn't yourself, can understand exactly what is required. For instance, in the JVC storyboard, the details of

Figure 2-2

Details of the JVC storyboard with the timing notations that were eventually assigned to each panel

the chamber walls had been deliberately kept vague to avoid any client disagreements. Before the storyboard could be realized, however, many more drawings detailing the walls and secret panels had to be executed.

When all the extra work of preparing a storyboard for actual animation is taken into account, you can add about a week to the schedule. This is not to say that everyone sits on their hands for five days while the creative director fills in necessary details. Plenty of work is still left to be done.

Break Down the Board

At this point, a key duty is to make a list of tasks and a schedule for their completion. Some tasks must be done first, because they lead to the completion of further tasks. For instance, a character must be designed before its choreography. But while the character's body is being designed, someone else can be starting the lip synching, that is, if the soundtrack is done!

For instance, here is the list of tasks for doing the JVC animation and the schedule we allotted for each task:

Task description	George	Jack	Days to execute	Deadline
Flesh out storyboard details	X		3	February 9–11
Write and approve script/dialog	X		5	February 10–15
Design exterior scene	X	X	2	February 9–10
Design interior hall		X	5	February 11–15
Acquire client art for moving stones	X		2	February 11–12
Design pharaoh's chamber		X	2	February 16–17
Design Egyptian character		X	3	February 18–20
Design professor character		X	3	February 21–23
Design pharaoh		X	2	February 23–24
Acquire music, effects, and voice talent	X		1	February 13
Choreograph Egyptian and Dr. Jones		X	5	Feb. 25–March 1
Record and mix voices track	X		1	February 17
Lip-synch Egyptian		X	2	March 2–3

Task description	George	Jack	Days to execute	Deadline
Lip-synch Dr. Jones		X	2	March 4–5
Lip-synch pharaoh		X	1	March 6
Render test pass in *standard definition* (SD) resolution		X	5	March 6–11
Test compositing of layers		X	1	March 11
Render corrections		X	1	March 12
Edit test and get client approval	X			March 13–15
Render final frames	X	X	10	March 16–26
Composite layers		X	1	March 27
Render and edit corrections	X	X	3	March 28–29
Record and mix master track	X	X	1	March 30
Conform to HD-D5	X		1	March 30–March 31

Allow Time to Render

Here's one consideration you should know right now, even though this consideration won't come up until much later. It is a common devil that rises its little ugly head toward the end of a project and has a nasty way of biting novice animators who think they have it all down. Take a look at the numbers in the previous minitable. You know how many frames you are going to create and you've figured out your board. You've planned how many characters and scenes you need and how long it's going to take you to design them. Have you considered how long it is going to take your computer to *render* these frames?

The rendering time is often the longest single block of time in any production schedule, especially if you are working on one computer. Early in your workflow, make at least one test frame of the most complex scene you will render. Record the time it takes to render this frame. Then multiply that time by the amount of frames in your production. In our JVC case, we were rendering highly complex frames with shadows, particle effects, and reflections. Because the end product was in HD format, each frame was four times the size of a typical standard definition frame. Our test frames required an average of 22 minutes per frame. Multiplying this time by the 1,800 frames necessary for a 1-minute production, we came up with a total rendering time of over 600 hours, almost 4 weeks of round-the-clock rendering!

The Rendering Farm

Knowing this, I realized I needed to build a rendering farm. This is a group of computers networked together and assigned to render simultaneously. LightWave, Maya, and 3ds max all support networked rendering. Using this capability, the software assigns various frames to be rendered by different computers, with the fastest computers getting the most frames. The software compiles the frames in numeric order on whatever hard drive(s) the animator selects. Although a network doesn't affect the rendering time of any specific frame, by rendering several frames at once, significant time is saved.

In order to evaluate the effectiveness of a render farm, it is necessary to render an entire scene of complex frames, sending enough frames to each computer to establish an average render time for the entire network. This process is easier if all your computers have the same specifications, but this is rarely the case in a small production company where both older and newer workstations, as well as the owners' laptops, are pressed into service.

Our render farm was composed of a clone dual Pentium II workstation, two Compaq EVO-8000W workstations, an IBM IntelliStation Z-Pro, and an IBM T-21 laptop. When optimized, the render farm allowed us to render a sequence of 120 frames in 72 minutes for an average per-frame rate of 6 minutes. This yielded a manageable estimated total render time of 180 hours, or about 8 days.

Notice how the schedule on pages 68–69 adds two days of buffer time into the schedule? Never pin a schedule down to exactly the time you estimate. The real world is full of "gotchas," Murphy's Law incidents, and everyday breakdowns. Whenever possible, add 10 percent of buffer time to each critical task.

The Animators' Tasks

Referring to the previous task list, let's use the JVC animation as a case history to explore the typical animation tasks that go into a complex project. Obviously, not every project will require all these tasks. Some animations are quite simple, requiring only a background and one moving object. However, by deconstructing a large project, you can see how several tasks can be combined, thereby seeing the individual steps as well as how they are contribute to the overall process.

By mastering the core tasks that compose a complex project, you can eventually bid on and produce large assignments. This method can also teach you to deconstruct even larger projects, facilitating the understanding of highly sophisticated animations, such as those used in leading feature films. Where possible, I will therefore explain the core tasks beyond the scope of this case study, so that you can learn how a company even larger than my own might have executed the same assignment.

Dividing up the Work

In large animation shops, each bit of work can be divided among many animators. For *Titanic,* one junior animator got to design and place hundreds of rivets. In the *Lord of the Rings* trilogy, a team of animators were assigned to capture the motion of one actor to provide action for the animated character, Golem.

Chances are, if you're going to run your own animation studio, however, you will have to wear many hats and divide a large variety of assignments to just a few workers. Such is the case in my shop, where two or three people produce major animations. In such studios, it's best to divide the work according to jobs that must be done at the same time.

Whereas a large shop will divide tasks based on seniority, experience, and talent, a small shop will divide tasks based primarily on the schedule, with the principal division usually being between picture and sound. Picture and sound are intricately linked in animation, but the jobs conveniently run parallel and allow for a good division of labor. Consider, for example, the necessity of **lip-synching** dialog, where an animator must make a character's lips move in synchronization with the sound. Obviously, the animator is going to need the sound to be produced before the lips can be animated. So, while one person is creating the character, the other can be casting and recording the character's voice.

Writing and Approving the Script

Writing a script for a client can always be a tricky endeavor. Although clients trust an animator with a wide latitude of freedom in the graphics department, they tend to compensate when it comes to the script. This is because most clients, at least industrial clients outside Hollywood or the major market networks, have no idea what goes into the making of an animation, but everyone knows (or thinks they know) how to write a good English sentence.

Writing a good English sentence, by the way, has little to do with writing a good animation script (professors take note). Famous commercials that have earned their clients billions of dollars in sales, from the 1950's "Winston tastes good, like a cigarette should!"[4] to the latest "He got game" flagrantly violate the rules of grammar, syntax, and diction. A good script therefore need not be perfect English, but above all a good script must sell.

You might ask what does the script of, say, a children's movie, sell? It sells the proposition that a donkey can actually talk and move the dramatic action of a 133-minute movie that made $267.7 million in domestic theater sales alone.[5] That's what it sells.

Writing a script for a client begins with understanding the client's bottom-line needs. What does the client want to sell and how can your script deliver the goods? Don't worry about the cool things you want to do with special effects or a new character. That's execution, and execution cannot begin until you and your client agree on bottom-line goals. Maybe as you're working with the client, in the back of your mind you can be thinking, "Wow, this is like some Viking going down into a cave to attack a dragon." But this moment might not be the time to express that thought to a client.

First, get the client to agree on some written **concept statements**, such as "Our task is daunting" or "We need a hero." Keep in mind that if these statements, which may only be subconscious expressions of your need to get to a dragon, are not working, you'd better be flexible or your client is quickly going to be feeling like he or she's been left out in the cold.

Once you've got some basic statements, it doesn't hurt to write them out on a white board or a large piece of paper. Once down, the statements can help propel you to a dramatic concept, dialog, setting, or time—all the elements that go into a good animation script.

---

[4]This slogan, from a 1953 cigarette campaign, was ranked in the top ten best by *Advertising Age* magazine. In the '50s, the population was aware of bad grammar and the brand received many complaints stating that the word "like" should have been "as." The sequel ad in response? "What do you want? Good grammar or good taste?" Today no argument is made. If you don't understand, "He got game," you're out of the target demographic. Grammar now functions as a smart-bomb element in advertising.

[5]I'm referring here to *Shrek,* 2001, DreamWorks SKG, David P. Allen, Effects Developer. Source: *Monday Times*, December 2001.

The Element Breakdown

Heath Vincent, the intern who came to us from Full Sail, showed me an exercise that he did for his professors. The exercise was to simply list the concept statements of an animation he was planning. You can do this really simply or you can get elaborate and make it look like a corporate organization chart, but either way, it's good procedure. Have a look at Heath's concept statement in Figure 2-3.

The next step in the element breakdown is to make a story summary. Again, keeping it simple, write out your story in simple declarative sentences. Figure 2-4 is Heath's result.

The next step in a major animation is breaking down both the physical and personality traits of the characters. You do this with a **character line chart**. Here the artist names the body parts that will be required. Simultaneously, another chart lists the personality attributes of the character. Later the two charts will merge to form a working model that expresses certain target emotions and actions (see Figures 2-5 and 2-6). You can see more about the process of designing a character and how Heath progressed with this animation in Chapter 7, "Rigging and Animating Characters."

The same technique can be used to build a good bottom-line script. The process merely gets you to break down a large, complex idea into simple elements. Maybe the client wants you to build up a brand name. You put the

Figure 2-3
Sample list of concept statements for an animation

Concept statements:

1. Good vs. evil:
2. Evil vs. good:
3. Battles are cool:
4. Be brave:
5. Be courageous:
6. Dragons are cool:
7. Knights are cool:
8. Simple stories work:
9. Being a perfectionist is painstaking:
10. I can't do math so I draw:
11. Warning: Neck hairs will stand:
12. Work is fun when you love it:
13. Focus on the answer not the problem:
14. We're all programmed:
15. One in the hand is worth two in the bush:
16. The grass is always greener on the other side:
17. I love computer animation:
18. Build the story around the feeling:
19. Work hard, computer animate hard:
20. Childhood inspiration:

Figure 2-4
Heath's story summary for the animation of a knight and dragon

Story Summary:

Company logo fade out, knight walking toward cave with torch at night. Fade in. Character enters cave cautiously. Storm going on outside. Dragon sneaks in through front entrance while Knight searches the blackness. Knight and Dragon face. Attack each other. Torch lands in water, scene ends.

Music?

Must . . . —chorus
　　　　—energy filled
　　　　—rain raising
　　　　—Dramatic

Figure 2-5
Heath's character line charts for the knight, one for appearance and one for personality

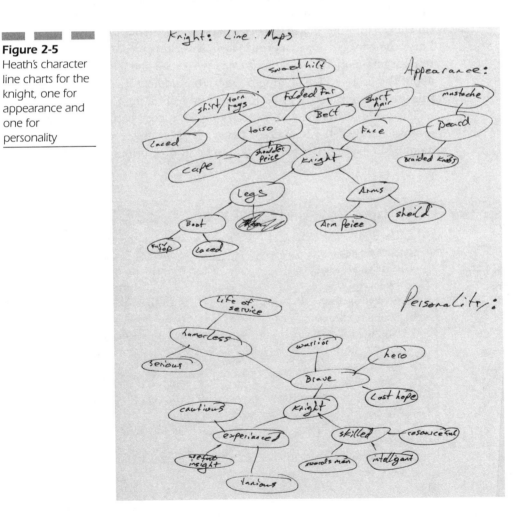

name in the center of the page and draw lines to words like "youthful" and "tasty." Then you can progress to actual statements that the brand can convey, such as "Doesn't this make you hungry?"

Milestones

Once you have a script the client likes, you have to stress that this represents a **milestone**. In contractual terms, a milestone is a point where a certain step is completed, or locked. Once the client has approved the script, hopefully in writing, explain that any changes to the script after this moment will entail significant extra costs. This is a bit scary, right? The client may back away for a moment saying, "Hey, wait. What if I decide I don't like a line or two afterward?"

What would happen? What if you've spent 10 hours lip-synching your character to the words your client now despises? If your client approved the script with the understanding defined previously, the client would have to pay the voice talent another session fee, the cost of re-recording and re-remixing the audio, and an extra 10 hours of work to change the character's lip movements, right? What if the client said, "A short red troll? No, I want a tall, green ogre?" More work. More money. And let's face it—*someone* has to pay.

If you don't want to be that someone, make it clear to your client that he's or she's passed a milestone. Get it in writing. And if the client isn't sure, back off and explore the options of rewriting and rescheduling, because locking a script is a little like getting married. It's expensive to change things once the guy in the robe says, "You may kiss the bride," and the parents have shelled out for a big reception at the plaza.

Design the Exterior Scene

An old comedian, Myron Cohen, once told a joke about a gigolo that was caught by a husband in his wife's closet. When the husband asked, "What the hell are you doing here?" The gigolo responded, "Everybody gotta be someplace."

The same goes for animation. Every animation gotta be someplace too. Even if you're making a simple logo fly into view, some "viewpoint" must be established for the logo to travel into (and to a lesser extent, some "viewpoint" from which it began). This view is called the **scene**. The scene is the central focus point of every animation program.

3-D scenes can be large, macrocosmic areas, such as a galactic space environment filled with stars, planets, and habitats or small, microcosmic areas, such as the interior of a human blood vessel or the molecules of a new compound. Even everyday environments can be complex, composed of various 3-D objects, such as furniture, trees, buildings, and roads.

When beginning an animation, it is wise to start with the scene, deciding first how big it will be. Our animation consisted of two scenes, an exterior desert scene showing a pyramid at night, and the interior hallways of the pyramid, leading up to the pharaoh's chamber.

The first scene was estimated to be about a mile square, whereas the second was about 100 feet long by 50 feet wide. We quickly realized that the exterior scene could be greatly reduced in complexity if the characters did not appear in the exterior but were introduced as already being inside the pyramid. By eliminating the 3-D characters, we eliminated the need to do any 3-D animation in the exterior scene. This meant we were free to create this scene using 2-D animation.

2-D scenes can be simple, still pictures that can be moved from left to right (**x-axis**), up and down (**y-axis**), or zoomed in and out (**z-axis**). These moves, however, are limited, because a 2-D image will reveal itself very quickly to be 2-D if the moves of the camera are not accompanied by similar changes in the geometry of the scene's objects. For example, in a 3-D scene, where the camera's view moves into the scene, the foreground objects must appear to separate to the left and right, exposing their sides. If the camera backs out of a scene, the human eye expects foreground objects to move into the scene from left and right while concealing their fonts. Therefore, the economical animator, using a 2-D scene, must be extremely careful with camera movement.

We used a software known as Bryce 5 ($169 from Corel [www.corel.com]) to create our Egyptian exterior. Bryce is a 3-D scenic animation program that enables the fast creation of realistic scenery, complete with mountains, stones, lakes, trees, and atmospherics. Once Jack had placed the pyramid, sand dunes, and moon, he produced a master 2-D frame that he loaded into 3D Max as a flat, planar background. Then he directed the camera to make the x-axis gently move while creating a faint backlight effect coming from the entrance of the pyramid. This was a perfect effect for the opening scene and, because it was only one frame in two dimensions, it rendered very quickly.

Flat-plane backgrounds, front projection, and/or rear projection capabilities are supported by nearly all 3-D animation programs. They are often used by animators to reduce complexity and render times. Planetary space backgrounds, forest scenes, and even complex office or home interiors can all be rendered onto a flat plane and incorporated in your scenes if elaborate camera moves or objects moving within the background are not required.

Design the Interior Hall

Because our characters were going to move through the inner halls of the pyramid, interact with the walls, and open a door to the inner chamber, we required a complete 3-D architectural design for the interior. We began by sketching a floor plan with specific lengths and heights. Animation programs enable you to assign a **scale** to a scene. This is done so that the animator can decide at the outset whether he or she is working in a space defined by nanometers, inches, kilometers, or, well, light-years.[6]

Keep in mind, however, that when working on a very large scale, the accuracy of the render is vulnerable to computer error. This occurs because the computer tends to round off numbers when calculating distances that are very large. This rounding, which is not consistent from object to object, can cause discernable drifting to occur between rendered points, producing mismatched mattes, distortions, and gaps in surfaces. When dealing in large scenes, it is often wise to scale down the dimensions.

Once the scale and floor plan are determined, you can actually start designing the scene. Exteriors usually begin with the ground, which may have topographical details, such as hills, holes, and even a body of water. The ground may be populated with natural and man-made objects, such as trees and buildings. The ground may also have surface attributes, such as a texture. Cement would be an example, or something more complicated, such as grass.

Walls that separate the interior from the exterior define interiors. Is the exterior seen through any port in the walls, such as a window? If so, you will need to construct the exterior scene or provide a flat-plane backdrop outside the window. Interiors may also have surface attributes such as glass, semi-gloss paint, or carpet.

As you can see from the pyramid example, our interior walls were textured to look like sandstone, with hieroglyphics engraved throughout. In addition, Jack decided to use torches as the source of light. This required not only the building of various interior details such as columns on which the torches would mount, but also realistic flame effects.

Here's a time-saving tip: Many novice animators (and a few careless experts) tend to overcreate their interiors and complex objects. Consider this question. If you are rendering a scene, composed of three rooms, and your action is taking place in only one room, is your computer rendering the unseen rooms? You bet it is. Even sophisticated programs and plug-ins that seek out unseen elements in a render order take time to work. In most cases, unseen elements in any particular scene can be eliminated with a huge savings in rendering time and storage space.

---

[6]Although I don't think any program actually specifies light-years as a dimension possibility. You can enter 180,000 miles one light-second as your stage width.

The implication of this tip is enormous. Early designers for seminal sci-fi feature films or TV series like *Babylon 5* and *The Last Starfighter*[7] quickly realized that by eliminating unseen details on space ships, rendering times could be cut to a fraction of their full-detail render times.

You can save even more time if you don't design unseen elements in the first place. Carefully consider your camera angles in advance (consult your storyboards!). If half the decorative balustrade of a stairway is going to be hidden by a character, don't build it.

Acquire the Client Art

In many animation projects, you will have to acquire art from your client or a third party. Such art usually includes logos and existing designs or photographs that are more economical to purchase than to shoot. Always exercise care in acquiring art, regardless of the source, to be sure that the rights of ownership do not conflict with your use. Be aware that using someone else's art can constitute a violation of the **copyright law**, which is punishable by a fine that can be as high as three times the amount you would have paid for the rights, plus court costs and lawyer's fees. Don't assume that art obtained from your client is delivered free of copyright entanglements. Often clients will send you art that was not properly cleared.

In our animation, the client supplied art for incorporation in the engraved hieroglyphics. Although most of the hieroglyphic designs were obtained by taking photographs directly off exhibits at local New York museums (this source is free and known as **public domain**), the special stones that the characters would press had to be derived from the history of JVC's technical research. One of the hieroglyphics was a picture of the famous Japanese engineer, whereas others were logos of significant JVC products like VHS and D-9. Each of these graphics was obtained by the client and transmitted to us under a contract attesting to the client's ownership. Proprietary logos, like JVC's "VHS" logo are closely regulated corporate assets with often complex rules governing their use in all media.

One of the difficult problems in character animation is the action of a character's feet on the floor. The problem is linking the character (or each

---

[7]Directed by Nick Castle in 1984, this was the first major motion picture to use computer-generated spaceships. A must see.

foot) temporarily to a particular point in a scene (in this case, the floor). When designing a scene, determine whether or not you need to show your character's feet (or the wheels of a car) touching the ground. If it's not necessary, a little bit of work may be eliminated.

With the exception of one scene where the Egyptian guide is standing still facing "the power," we eliminated the extra work of linking feet to the floor by never showing the characters' feet while they were walking. The *effect* of walking, however, is greatly enhanced by the **sound effects** of the characters' feet on gravel. Never underestimate the power of sound in animation. That's why the *Enterprise* in *Star Trek* always makes a whooshing sound, even though in a vacuum, no sound would be heard.

Designing the pharaoh's chamber entailed one surface attribute that was not included elsewhere in the animation, and this is one you might want to use someday: the incorporation of moving images within a scene, such as a movie. In our case, we used a planar, projected image sequence to create the illusion of a movie projection on the wall of the pharaoh's chamber. Moving image sequences can easily be imported into most professional animation software. These sequences can be frames of video or another animation. The sequence can be viewed on any surface—planar, convex, or concave—within a scene. You can use this technique in the same way you used a still frame as a backdrop, only this version will have moving images. You can also use it to create moving images on a TV screen, within a crystal ball, or reflected off the surface of a rippling pond.

Design the Characters

Someday a computer program and a peripheral device may be invented that make it easier to design a character straight into the computer without any preliminary steps. The peripheral device might be something that attaches to your hands and lets you sculpt virtual clay seen through a pair of 3-D goggles. Certainly, the new Cintiq line of monitor tablets from Wacom (www.wacom.com) enable a pen device to draw directly on a flat-panel monitor. This is an important step in the right direction, allowing an animator to sketch electronically (see Figure 2-7).

For now, however, the best way to start creating a character is to draw him, her, or it on a sheet of paper or to sculpt the character with some material that will solidify. Pat Starace, a California-based animator with a sculpturing background, prefers to use **Super Sculpey** modeling clay (www.sculpey.com). He creates the character around a wire form, pushing

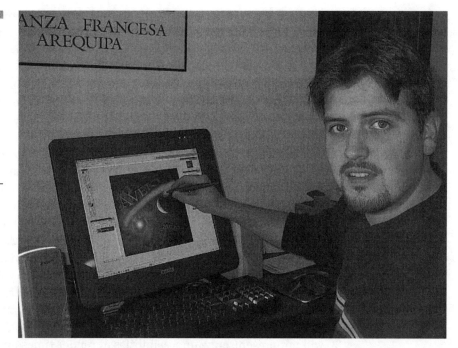

Figure 2-7
Animator Tom Maloney demonstrates our Cintiq 18SX monitor tablet from Wacom (list price, $3,499). A smaller version, the 15X, sells for $1,899.

and pulling using his bare hands and a few basic tools until he has just the look he wants. Super Sculpey can then be permanently hardened in a home oven at just 300 degrees for 15 to 20 minutes. Often, he will make several models of the same character, each with different expressions or body poses, matching the key frames of his storyboard.

When the characters are just right, Pat takes a sharp pencil and draws a grid over the surface of the character. Then he uses a pointing device, such as the **Microscribe G2** ($3,495) from Immersion Corporation (www.immersion.com), connected directly to his 3-D animation program (Microscribe products are compatible to LightWave, Maya, and 3ds max). The program records every press of the pointer's tip in virtual 3-D space. As Pat presses the tip on every intersection of the character's grid, the 3-D animation software builds the surface of the character onscreen.

Major feature-film animators use this same technique when creating creatures that have never been designed before. Some years ago, I interviewed the creators of the werewolf in *American Werewolf in London*. Located around the studio were white plaster models of the werewolf in various stages of transformation from human to monster. Each model bore an intricate grid of lines across its surface. Obviously, the big guys were using the same techniques as their freelance compatriot Pat Starace. Today faster input can be achieved using laser-based capture tools, but the cost of these tools is still high for the entry-level animator.

In creating the three characters for our JVC animation, we started simple. Jack made pencil sketches of the characters from various angles, offering me, the producer, an assortment of choices, such as noses, hands, postures, clothing, and hair.

Hair and clothing are still major concerns in animation art because they require complex motion to appear realistic. While a character's arms and legs move, they usually do so according to a rather simple set of rules (except for an octopus). An human elbow joint, for instance, can move about 175 degrees on only one axis in one dimension. The shoulder is a bit more complex, with three dimensions of rotation. But hair? A single strand of hair can move like a snake, with infinite axes, rotating nearly 360 degrees in three dimensions. Clothing gets even more complex when you consider that the surface of clothing is like a patchwork of hair strands.

It takes a wise producer and an animator with the strategic skills of a platoon commander to plan the design of a character on day 1 that by day 30 will not present huge impossibilities of render times and computational difficulties. As an example, Jack and I originally planned the Egyptian character to have a flowing robe. Using Digimation's **Stitch 1.1** (www. digimation.com), a program specifically designed for creating flowing cloth in 3ds max,[8] Jack showed me some exciting results early in the character design process. By the time we were closing in on our deadline, the robe, which was taking an extra 10 minutes per frame to render, had to go. Had we known this in the beginning, we could have used the time spent on the robe to create some other cool feature (see Figure 2-8).

Hair is another tricky element that can eat up time. That's why you see so many novice show reels featuring bald characters or characters with short, moussed hair. Nobody wants to wait for hair to render. As you can see from our humble example, the Egyptian is bald, Dr. Jones has a crew cut, and the pharaoh is wearing a headdress.

---

[8]Other tools you might use to create cloth in 3ds max include **Reactor** and **ClothReyes** from Infographica. In Maya, cloth tools include **Maya Cloth** (www.aliaswavefront.com/en/WhatWeDo/maya/see/solutions/mu_cloth.shtml) and **Syflex** (www.syflex.biz). In Light-Wave, cloth plug-ins include **Motion Designer**, which was bundled in all versions of Light-Wave from version 6 on, and **Cinema 4D** from Maxon (www.maxon.de).

Many tools exist for designing character heads and bodies. **Famous 3D proFACE** (www.metamotion.com, **$1,995**), Curious Labs' **Poser** (www.curiouslabs.com, **$319**), and **Credo Interactive's Life Forms Studio** (www.credo-interactive.com, $295) are excellent tools compatible with a wide range of animation programs. Searching the Internet will result in lists of libraries with a wide range of existing (sometimes free) characters and elements from which to choose.

Of course, you can design from scratch as well. Unless you are on a really tight deadline or working with a standard human figure, in which case it would be stupid to redesign the body from scratch, it's esthetically preferred to design a character from scratch. It doesn't take that long, the results are uniquely yours, and when you show your work to a real pro, he won't be thinking, "Oh yeah, there's the Poser4 Victoria model with a different nose."

If you're in a real bind, you can also choose to buy your characters from stock 3-D object houses such as **Digimation** (www.digimation.com), who exclusively represent the 3-D model catalog once developed by **Viewpoint Data Labs** (www.viewpoint.com). You can also download free characters from any one of several web sites that seem to come and go with great frequency. Simply type "3-D character object X" (where X would be the type, such as "fat man") into your search engine and see what comes up.

A while back, Jack and I were working on a logo for Compaq Computers (now HP) and we needed a cheetah. Viewpoint had a nice one that only needed a bit of tweaking to make it work for us, but a cheetah is a body type that exists for real in nature. Only one correct cheetah exists. That's different from creating a character. A character is a personal artistic expression that should not, in my opinion, be obtained off the shelf.

I should also note here that our time estimate for designing the characters, which you'll note from the task breakdown, was pitifully short. A good character animator can create a full human character from scratch in about a week. **Rigging** the character, that is, putting in the bones, joints, and controls that enable the character to be choreographed, takes another week. Actual choreography can take a few days or more, depending on how much action is required. The time we allowed ourselves meant that Jack wasn't going to get much sleep, but he knew this going into the project. Plus, he was working with some character types he had already crafted months earlier in his spare time.

Acquiring Music, Sound Effects, and Voice Talent

You have three choices when it comes to obtaining music and sound effects. The first two are legal. You can make your own or you can buy the rights to use someone else's property. I caution you severely when it comes to the third possibility: acquiring audio elements without the appropriate copyright permission.

As noted previously, the penalties, if you are caught, range up to three times the cost of the rights had you obtained them legally, plus your lawyer's fees, maybe the copyright owner's lawyer's fees (that's up to the judge), and the court costs. During the two or three years in which your case winds its way through the legal system, your production most likely will be legally seized and you won't be able to use it.

It's easy for the owner of the music to find out about your production, and entire organizations have been established to assist the owner in making your life hell. It's a bit harder to apprehend copyright infringement involving sound effects, especially if the effect is derived from nature and buried in the tracks of a production. But why risk the hassle?

Granted, the temptations are there. Vast sites exist on the Internet, full of sound effects, some brazenly and obviously pirated from famous feature films. Obviously, it would be inefficient for Lucasfilm to search down and sue every 14-year old webmaster with a clip from *Star Wars*, but if you boosted James Earl Jones's voice saying, "I'm your father, Luke," for your major-market animated commercial for Nike, you could bet your lightsaber that a **cease and desist** letter from Lucasfilm's legal department would be in your mailbox in short order. And don't bet on Verizon taking a backseat with their rights to Mr. Jones's voice either.

Music Production Software If you're inclined to make your own music, several great products are on the market and can turn the computer you use for animation into a fine synthesizer-sequencer. **Sonar XL** ($599) from **Cake Walk** (www.cakewalk.com) and **Cubase SX** ($600) from **Steinberg** (now a division of **Pinnacle Systems**, www.pinnaclesys.com) are music composition programs that enable you to create your work on a computer and then print out sheet music (**charts**) if you want to go into a studio with live musicians and record your opus.

Reason ($399) from **Propellerhead** (www.Propellerhead.se) offers a complete, modular sound studio including dozens of "rack-mountable" units, such as a synthesizer, mixer, drum machine, and mixer to provide a user with a complete music creation solution. **ACID Pro** ($350) from **Sonic Foundry** (www.sonicfoundry.com) enables a novice with little musical background to crank out a pretty good composition, multi-tracked and output to any convenient computer media file using simple, prerecorded loop elements supplied on an ever-growing list of inexpensive, **royalty-free** CDs.

Simpler programs such as **Music Creator** from Cakewalk ($49.95), **Cubasis VST** from Steinberg ($99), and **Music Master Works** ($39) from Aspire Software (www.musicmasterworks.com) are excellent if you just want a low-cost solution or introductory music creation software on your workstation. **Music Master Works** incidentally features a clever "voice-to-**MIDI**" (for **Musical Instrument Digital Interface**) recorder that enables you to sing into your computer and produce printable music charts and MIDI sequences that play in a variety of instruments at the click of your mouse (see Figure 2-9).

Stock Music One of the easiest and sometimes most economical ways of obtaining music for your animations is to buy your cuts from a **stock**

Figure 2-9
A sample interface from Music Master Works from Aspire Software, which enables a user to sing into a computer's microphone to produce MIDI passages and printable charts

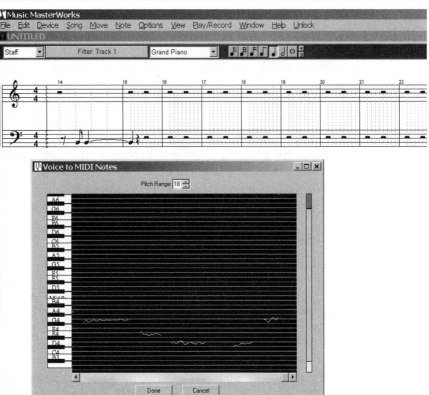

music house. A stock music house acquires thousands of cuts of music with full rights for all media worldwide. Then it issues these cuts over secure Internet connections or on CDs to producers like you. You buy the CDs or pay a fee to access the music on the Net. Each time you use a clip of music for a production, you pay a fee to the stock music house.

Fees are usually based on one of three methods. A **needle drop**, the first, means you pay for each time the "needle" of your phonograph touches the vinyl. For those of you who don't remember vinyl, consider the needle drop to be any cut of any music for whatever length. If you use one piece of music and extend it with two edits, that's three needle drops.

The second method of billing for stock music is to pay a **production fee,** which allows you to use the entire library and as many clips as you want for one production. This fee is usually about the same cost as three needle drops.

The third method is based on a **yearly license fee**, which licenses you to use defined portions of the stock company's library or the entire library without regard to the length, number of needle drops, or number of productions. Fees are also adjusted to account for the kinds of audiences for which you will produce, based on the needle-drop rates for each type of audience, for instance, nonbroadcast industrial productions cost about $75 per needle drop at the low end and national television commercials cost over $300 per drop at the high end. Some of the best-known stock music houses are **OmniMusic, Killer Tracks, ProMusic,** and **FirstCom/ MusicHouse**.

Sound Effects   One of my first jobs in the film industry was to go to a cemetery in Princeton, New Jersey and record birds chirping for a documentary producer. It's fun to gather your own sound effects and store them on CDs. If you haven't got the inclination, great sound effect libraries are available on CD. One firm, **Sound Ideas** (www.sound-ideas.com), offers several series of CDs, ranging from a Super Sampler (1 CD for $129) up to their master collection, Series 6000, The General (40 CDs for $1,495).

As mentioned before, millions of sound effects are available on the Internet, but before you download and use them, I suggest you verify that the source site owns the effects and gives you permission to use them. I would include some sites here, but I've found that sites and their policies change so frequently that the information would become obsolete too quickly.

When it came to producing our JVC animation, I had little time to play around looking for the right music and effects, even though I have a complete stock music and effects library in house. At such times, it pays to subcontract all or a portion of the audio. Here's how this can work.

Find a reliable sound studio owned by a technician-artist who really knows how to craft a good soundtrack. This is quickly becoming a specialty called **sound design**. My two favorite sound designers in New York are **Howard Schwartz** (www.hsrnewyork.com) and **Jeremy Goldsmith** of Tabby Sound (www.tabbysound.com). At the time of this production, Jeremy's smaller studio was limited to two-track mastering, whereas Howard's shop is replete with all the latest Surround Sound and Dolby 5.1 mastering capabilities. At first, JVC wanted a 5.1 audio track, but later, when the client determined that their exhibit booth would not support an HD playback with 5.1, the specifications were lowered to simple stereo, so Jeremy got the subcontract.

These details are provided to illustrate how a client's specifications may change and how your production must adapt within a certain unpredictable range of variables. You need to keep in daily contact with your client to know about these changes in time. Clients make lots of decisions during a project that they may not immediately equate with a need-to-know message to you. It's your responsibility to find out. Once you find out, it's your responsibility to assess the impact on your production. Does the change increase your cost and require an overage warning? If so, you must tell your client, "This goes beyond our contract and will cost you more money." The change may *reduce* your cost, in which case, if your contract is for a fixed price, you get to keep the difference.

Knowing that I was going to record and mix my soundtrack at Tabby Sound, I emailed Jeremy my script and storyboard a few weeks ahead of the record date, highlighting the sound effects and music notations. Then I left it up to Jeremy to collect the elements, which I checked a week later. A good sound designer will add his own level of brilliance to a project, and Jeremy was no slouch. He figured out that I was going to need "feet on gravel" sound effects wherever the characters were walking and some Egyptian chants to set the mood for the opening exterior scene. I hadn't considered these and was delighted with the sound designer's input. Lesson to be learned: If you want to be a great producer, surround yourself with great talent.

Casting Talent   After a week of listing your production company in the phone directory, you should be getting calls from people with fine-sounding voices, offering their services as voice talent. Ask them to send you a cassette or CD of their work and store these for the day when you need to cast a job, because on *that* day, no one will call you offering their services.

You can also call talent agents, theater clubs, universities (professors often have great voices), churches, and synagogues (New York's Anglican bishop, Herb Gross, is a dead ringer sound-alike for James Earl Jones). Sound-alikes, by the way, are legal as long as the talent does not identify him- or herself as the real person or as a character that the real person is known for.

When casting voice-over talent, you're allowed a liberal amount of time to see if the person can create the voice you want and, if necessary, lip-synch to a video playback of the character. In our case, we planned to synch the character to a prerecorded voice, but when working with high-cost celebrity talent, or for other logistical reasons, you may have to match the voice to the picture. In such a case, it pays to know if your talent actually has this skill, because it isn't that easy and a studio equipped to record lip synching (also known as **looping** or **automatic dialog replacement [ADR]**) costs about $350 per hour.

I like to find about three actors for each part I want to cast. Over the years, I've found some voices that I love and who have become my favorites, but I try to have each casting session include at least one newcomer for each category. Never close your doors to any new talent. Your job as a producer is to constantly meet new talent and keep your roster changing for the better. Lesson to be learned: If you close your door, you are not doing your job.

Here's how a casting can go. For Dr. Jones, I listened to about 50 tapes of male voices that I had in my library. From these, I picked my five favorites and then played them for Jack. Jack and I selected the best three and then we asked them to meet us at the sound studio.

The character of the Egyptian was a bit tougher. Obviously, the demand for Egyptian voice-over talent in New York isn't so great that Egyptian actors would be sending tapes to small animation studios. The same goes for character actors who specialize in crafting an accurate accent for any nationality or region. At first, I started calling the few character voice talents in my library and asked them to try an Egyptian accent over the phone. This produced some pretty hilarious results.

"You want Omar Sharif or Sabu?"

"Who's Sabu?"

"This voice. This is Sabu."

"Sabu to you."

"This is good Sabu."

"Sounds more like Dr. Who."

My wife, Maria, who runs the language translation part of our business, suggested calling the Egyptian consulate and the United Nations to find a genuine Egyptian. She was really lucky. An Egyptian translator named Fawzi Abdeltawab visited us at our studio and read the script out loud. He was perfect for the part. While reading, he politely asked if he could address an issue regarding the script. I welcomed his comment.

"You have Dr. Jones addressing the Egyptian by the name of Hakim. With all respect, I must tell you, Hakim is not an Egyptian name."

I was seriously concerned. Our efforts at using a real Egyptian instead of a professional impersonator derived from my desire for authenticity, so I asked Fawzi what name he would suggest.

"Faraq would be better."

"Faraq?"

"Yes. It means 'the one who searches.'"

"Faraq it is." Lesson to be learned: Listen to people who know what they are talking about.

Having selected our chosen talent for the voices, we recorded the dialog and edited a final voice track. Later, after all the animation was completed, we planned to make a final mix to add the music and effects. For now, we had the only element of sound—voices for lip synching—that was required to complete the animation. We locked in our appointment for finishing the mix, approved the elements that Jeremy, the sound designer, had assembled, and returned to the studio to continue working.

Choreographing the Egyptian and the Professor

The first step in choreographing a character is to decide which parts of the character actually require movement. For instance, the pharaoh was going to be sitting on a throne, eating popcorn, and laughing. Therefore, once the pharaoh was positioned on the throne, all we had to move was one hand, his head, and his face. After some thought, Jack suggested that if one hand was holding the popcorn box and one was already in the popcorn, none of the hands actually needed to move. That simple thought saved us a day of labor. Lesson to be learned: Think simple.

Dr. Jones and the Egyptian, however, were going to have to walk and talk and push stone slabs on the walls of the pyramid. These characters were going to require a full set of human bones and controls, what animators call **rigging**.

Any character that moves around on a stage will require some data relating to the character's structure that defines for the computer how the animator intends the character to move. This data is usually expressed as hierarchical, articulated structures that enable static positioning and animating of any deformable object.

Hierarchy, by the way, defines how one set of control data will be slaved or slave another set. For instance, an ankle joint is slaved to the lower leg bone so that when the bone moves the ankle will follow along, all data that describes the ankle is, therefore, slaved to the data that defines the lower leg bone. Before hierarchy was developed, the animator had first to move the foot, then reconnect the leg, and so on. Aren't you glad you weren't born earlier?

This data (referred to in each program as **bones**) is usually displayed in a software interface as some overlaid graphic image, most often forming a line or shape that resides within the portion of the character you want to control. In some programs, you can actually grab and manipulate the control element, changing such attributes as its location, motion limitations, strength, and hierarchy.

Jack took the 3-D object collection that composed each of the two characters and installed the bones and joints that would cause the objects to move in predictable ways to emulate real, human movement. Once rigged, each moveable part (a leg, foot, toe, or whatever) can be assigned a **key frame**.

Key Frames In animation, frames define the dimension of time. Your animation might start at frame 0, for instance, and proceed to frame 30 a second later. If you want an object to move in that second, you have to give that object a starting point in the first frame, a key frame, and an end-point in the last frame, also a key frame. The computer then knows where the object is starting and where it is ending. The computer then figures out all the in-between movements of the object and draws them for you in frames 1 through 29. To make the Dr. Jones character walk, for example, Jack had to position each of the character's feet in a starting key frame. Then after pulling the joints and bones of the character to another position, say, the first foot forward, he created another key frame.

F-Curves, Envelopes, and Time Splines As you may know, the real world is perceived through our senses as a vast array of bioelectric impulses to our brains. These impulses, and many of the natural world phenomena that cause them, are analog. That is to say, they are composed of irregular, infinitesimal fluctuations of values and random noise.

The digital world, in which animators attempt to imitate real-world phenomena, is composed of periodic, finite values devoid of random noise. As an example, in order to smile, a human face employs a certain number of muscles, billions of neurons, and trillions of cells. No computer can yet emulate this level of analog complexity. The most sophisticated digital rendition of a human smile will emulate only a minute fraction of the real-world in order to attain a result that will fool the human mind in a theatrical setting (in my opinion, this is yet to be done).

The problem animators have in attaining realism is based on the keyframe. If a keyframe is the digital repository of all the parameters of a human smile at a given instant in time, what about the frame before and after the keyframe. What does it contain? It contains the computer's *interpolated* values from those of the nearest keyframe to those of the next (or previous) keyframe. For instance, if the tip of the nose is at point "A" in frame 30 and at point "B" in frame 40, the computer—unless you specify otherwise—is going to interpolate an *equal* change of position for each frame from 31 to 39. Often, the computer interpolates unrealistically, producing results that are obviously wrong. In most cases, it is impractical to tweak every frame of an animation, essentially making it a keyframe. We have to rely on the computer. But how can we make the computer avoid those obviously machine-made interpolations, such as when a curve ball lands in the dirt rather than sliding gracefully along a curved path to the catcher's mitt?

In such cases, you must employ tools such as **function curves**, **envelopes**, **motion graphs**, and **time splines**. These devices provide graphic representations of the calculations between key frames. For instance, using these devices (they are known by various names in different software), you can see the movement of the curve ball from the pitcher's fingers to the catcher's glove as a ramped graph line expressing its position in x, y, and z for every frame. If the computer produces an interpolation that is too machine-like, say a sloped, straight line, you can bend the line to add real world abberations to the movement. One common machine-made effect is seen as a sharp corner in the function curve, representing a strictly mathematical interpretation of a change in motion. If a corner appears in your function curve, you can see it as you can apply various tools to that curve to round out the keyframe and produce more realistic results. We'll delve into these controls later in Chapter 4, "The Mechanics of Three Leading Animation Programs," and Chapter 7.

The Timeline   Most animation programs feature a horizontal timeline, whose length represents the length of the animation. The timeline usually includes a superimposed grid of vertical and horizontal lines. The vertical lines divide the timeline into frames, seconds, minutes, and so on. The horizontal lines allow actions to be placed on the timeline. These actions, usually shown as a colored or shaded horizontal ribbon, can be 3-D objects with at least starting and ending keyframes (these two keyframes denote the ends of the ribbon). The ribbons can also be special effects, lights, or other features within the animation, each with keyframes. You can control these elements, their starting and ending frames, and their appearance and disappearance in your animation by sliding these elements or their beginning and end points left and right.

The Camera   Another important element of choreography is the camera, through which the computer program will render the scene that the viewer will see. The camera can also be choreographed with its own keyframes. Most animation programs feature the usual repertoire of **camera moves**, such as **panning** (left and right turn), **tilting** (up and down), **tracking** (in and out of the scene), **dollying** (left and right), and **ped** or **pedestal** (up and down elevation). Some programs also enable **camera lens adjustments**, such as **zooming**, **f-stops**, and **aspect ratios** (the height-to-width relationship of the frame). Whenever a move or lens adjustment is initiated, it must be keyframed with a start frame, any intermediate steps, and the end key frame.

Lights   Lights can also be choreographed in a scene, although they usually remain static. Obviously, a static object will have one key frame to start and no other keyframes unless it moves or changes its parameters, such as light color, focus, and intensity. Any parameter of the light that may be defined is usually keyframeable, which means that the parameter can be made to change predictably over time.

Coordinates   Here's where your high school geometry teacher gets her revenge. Remember the famous Cartesian plane? It's *ba-ack*. All movement within the 3-D environment is expressed in x (left and right), y (up and down) and z (in and out) coordinates. If you move the camera, a light, or any object, you must do so in terms of x, y, or z amounts within a space that is measured in the same three dimensions.

Similarly, many programs offer views of your work based on the Cartesian plane. For instance, you may be offered a top view that is also identified as the z–x axis view, because it allows you to see the stage in terms of these two dimensions as you peer down on the y-axis. Movements of objects are also referred to both on the stage and in numeric readouts as x, y, and z coordinates.

After rigging his characters, Jack could move them about the stage like puppets. For instance, he could place Dr. Jones at a beginning point in his walk down the hallway. Jack could assign the starting position for Dr. Jones on the timeline and store keyframe 1 to record all of Dr. Jones's surface, bone, and joint elements at this particular starting point. Then Jack could move the left foot of Dr. Jones forward to begin the walk sequence. Because all the joints and bones of the character's foot were chained by hierarchies to the rest of the bones and joints, when Jack moves the foot, the rest of the character follows along, but not very realistically.

When real people walk, their shoulders move, weight is shifted forward, hands swing, and the head bobs. These things don't happen by accident in animation. Each movement needs starting and ending key frames and often several key frames in between to create a realistic character action. After moving the left foot one step forward, Jack could define keyframe 2 about a second forward on the timeline. That keyframe would record the second position of the foot and the computer could, from keyframes 1 and 2, generate all the action frames in between. Still more work needs to be done.

If Jack asked the computer to generate the frames between his two keyframes at this point, the foot would remain in the same flat horizontal position it had in the first two keyframes , thereby looking like Pinocchio. In order to give the walk realism, Jack would have to give the foot a bit of downward rotation about midway through the step. He would record a new left-foot keyframe 2, and the computer would automatically renumber the old left-foot keyframe 2 as 3. Then, when Jack asks the computer to generate the in-between frames from keyframe 1 to 3, the computer would have to take into consideration the new downward rotation midway through the step, allowing the toes to point a bit downward, like they do in a real human. After keyframe 2, the computer would start rotating the foot back to horizontal so it can land on the ground flatly.

The same intricate process of keyframing must be carried out for all the parts of the character that move during the walk sequence. If the head, for instance, must have a slight bob downward with each step, Jack would have to add a head keyframe in midstep, storing the lowest point of head-bob and then another head keyframe at the end of the step to bring the head back up. The process of getting both characters to move throughout the pyramid and finally meet the pharaoh in the final scene took about two weeks of work, with very little sleep.

Motion Capture   Several companies make a useful device known by the generic term **motion capture system**. Simply put, a motion capture system enables a computer to record the position, orientation, and motions of a live subject, such as an actor, and assign them to a 3-D animated character. Perhaps you have seen such systems being used in documentaries about animated feature film projects. Several types are available, including the wearable sensor from **Immersion** (www.immersion.com), the electromagnetic **Polhemus Fastrak** (www.polhemus.com), the infrared, acoustic **5DT** (www.5dt.com), Metamotion's **Gypsy3** (www.metamotion. com), and the electrostatic and electro-optical **Vicon** (www.vicon.com).

Each of these systems employs a method of achieving three goals: identifying each moveable element of the subject's anatomy, tracking the position in a 42 space of each element, and reporting all position and motion data to a computer.

All these systems employ a high degree of research, development, proprietary devices, and precision, making them all quite expensive (mostly over $5000). Each type of system has its pros and cons, and many expert animators feel that the perfect motion capture system is yet to be developed. Results vary from system to system, and none provides perfect data acquisition, such that little time is needed to tweak the capture results for

Figure 2-10

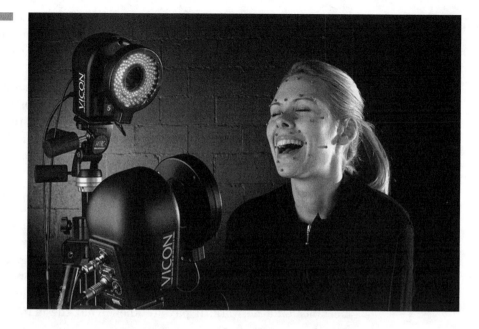

final application. Some animators, the purists, swear that they can achieve equally fast results by hand-coding the keyframes. Certainly, if you are planning a large project such as *The Mummy* (which used the Vicon system) or the *Lord of the Rings* saga, you'll probably opt for a mocap system, but as an animator just starting out, you may opt to perfect your keyframing skills by hand (see Figure 2-10).

Building a Render Farm

This book is designed to take you from having no knowledge of animation to the stage where you are ready to open and successfully run your own animation studio. During this journey, you may not have much of an opportunity to work with mass-production hardware, but I urge you to use whatever opportunity may come your way to gain experience in computer maintenance and network management.

Get to Know Computer Hardware   Typically, when you start your own studio, you will be working alone with one computer. This isn't a good place to start learning about networking. Obtain network experience in your internship and entry-level job days. If your internship or first job is coloring or accounting, take your lunch break to go chat with the person running the computer network. Nearly every medium-sized business has a network. Even if you're working in a law firm or a garden supply center, look for the network and learn it. You will not be wasting your time.

Knowledge of your computer will assist you in maintaining the system and avoiding down time. Further knowledge of your computer will allow you to diagnose and perhaps repair your system without relying on outside vendors or an employee who may not be available when disaster strikes. Finally, networking knowledge will allow you to assemble powerful **render farms** whenever you need to execute a large project.

Network Rendering Many 3-D animation programs offer ways to render frames over a network. You should determine, prior to choosing a program, what the manufacturer's policy is regarding network rendering. Some manufacturers do not support networked rendering. Others want you to pay an extra fee for each machine connected to the network. Still others include network rendering for free for a limited or unlimited number of machines on a network. It pays to know what the manufacturer of your favorite software offers.

Maya Unlimited enables an unlimited number of rendering computers to be connected to a network, but a minimal version of Maya must at least be installed and licensed for each machine. 3ds max allows unlimited networked computers, but the maximum you can run without a dedicated server is 10 and each must be installed with the included Netrender software.[9] LightWave's included ScreamerNet software will enable any amount of rendering computers to be connected to one licensed workstation.[10]

What Is a Render Farm? A render farm is a group of two or more computers networked (sometimes temporarily) in such a way that your workstation can delegate the rendering of animation frames to this group. Obviously, a render farm will render finished frames faster than a workstation.

A render farm can be composed of almost any computer in your studio, and if you are familiar with the methods of networking, the farm can be

---

[9]Additional machines can be added beyond 10 using a server. When doing so, it is wise to use an additional mirrored server for configurations over 25 machines. But if you're doing this, you're way beyond the scope of this book!

[10]Limitations within the operating system may apply to render farms using LightWave ScreamerNet. For instance, Windows NT has a limit of 10 rendering computers. Consult with the vendor before investing in hardware.

built in a day or less, employed for the task at hand, and then disassembled so the computers can be returned to their original function. In many cases, the render farm can be left in place, with the computers returned to their previous function while still networked and ready to be used again.

Making a render farm is easy. If you have two or more computers at home, even if one is an Apple and the other a Windows model, you can network them together and practice building a render farm. When you're done rendering, your family will be able to enjoy the network and share files and devices like hard drives and printers.

In fact, in our shop we make the assembly of a render farm one of the first things an intern will do. It stretches the experience level, it's usually pretty harmless (if the network doesn't work, it doesn't work, but all the computers are still operational), and when done, it makes the intern feel great about having done something important. Making a good render farm, however, is a tad more difficult and requires some planning in selecting the right computers.

Deciding Which Computers to Network   The criteria for selecting the computers for a render farm are based on the relative speed of each computer and the hard drive storage of each. Your animation-optimized computer is probably the fastest one in your shop. Make a list of all the others, including your laptop. List the computers' makes and model numbers, the **Random Access Memory (RAM)** capacities, the speed of the **central processing unit (CPU)** and the available hard drive space. The CPU is usually expressed in **hertz** or **Hz**, such as 700 MHz, which means 700 **megahertz** or 700 million cycles per second, or 2.2 **gigahertz** (GHz), which means 2.2 billion cycles per second).

The most successful render farms are those that group several computers with similar CPU speeds. If, for instance, you have two computers with a 700 MHz Pentium IV chip and one with a 286 MHz, pre-Pentium CPU, it would be wise not to include the slower machine in the farm, because the other computers would be rendering several frames while the slower was rendering one, thereby slowing up the assemblage of frames.

A well-matched network consists of computers with CPUs that are within about 50 percent of each other's processing speed. It is also important for each computer to have adequate hard drive storage and RAM capacity, because this is where the intermediate steps of rendering and the final rendered frames will be stored until they are fed to the artist's hard drive. If the RAM or hard drive space of a fast computer is not adequate, consider adding RAM chips, freeing hard drive space, or installing an extra drive.

Another important issue in matching CPUs is to match brands. I've noticed significant rendering errors in networks that employ both Intel and AMD CPUs, because the two brands handle their mathematics in slightly different ways. The result is that subtle differences occur between frames if a sequence is rendered across both brands. If you must employ mismatched CPUs in your rendering farm, try to segregate entire sequences to one type of CPU. The differences in rendered frames can often be absorbed by a scene cut, but not frame by frame.

The scope of this book does not include specific instructions for setting up a network, but I encourage you to consult your *operating system's* (OS's) instructions. The latest versions of the most popular OS software have made networking rather simple, especially when building a **peer-to-peer network** (a network without a central server). Windows XP, for instance, takes only a few minutes to link two computers, even if you're doing it for the first time, whereas the older Windows NT 4.0 takes an hour or more.

When you've got your render farm working, you should be able to open a file on any drive on the network from any computer on the network, and save that file onto any drive on the network. That test completed, you are ready to configure your animation software to use the network to mass-render frames.

Once I had the render farm built and tested, Jack could start pumping humongous amounts of frames through it, which greatly reduced the time he needed to test animation sequences. Animation is really a process of iterations. The more time you have, the more trials and errors (iterations) you can test and the better your end result. If you have only one animation machine, you have to wait to see the results of each frame. When you want to check how several frames look played together, such as to test how a character looks walking in the rain, you have to wait until enough frames are rendered to evaluate your results. Therefore, in producing quality animation, the more frames you can render at one time, and the faster each can render, the more successful you will be.

Lip-Synching Characters

Synchronizing the lip movements of a character to match the words of a recorded voice track is called **lip synching**. Lip-synching a character is a lot like choreographing, but on a smaller scale. In fact, motion capture systems are designed for capturing facial and lip movements of an actor in real time.

In the same way that live motion capture is not used unless the project is for a regular, repetitive, or large project, lip synching by hand is preferred by most animators who only require the technique occasionally or who need to lip-synch several characters in short sequences. As a beginning animator, the target audience of this book, chances are your first experiences with lip synching will not be enhanced with a live motion capture system. Don't feel bad. After more than 10 years' of work in the business and just as many years writing about it, I haven't had a chance to try motion capture lip synch either.

After you get the hang of it, lip synching by hand can be fun and you can get about a sentence a day done on a good software program. The technique begins by importing the voice of the character into your animation program as an **.mp3** or **.wav file** (the two most popular audio file formats).

Many animation programs feature the capability to import and scrub through sound using a tool that mimics a ***nonlinear editing*** (**NLE**) program or ***video cassette recorder*** (**VCR**). First, you link the audio track to the animation software's timeline so that when you scrub forward and backward on the timeline, the audio plays along. Then you place the first frame of your character's lip movement to the first frame of the voice track. Now scrub a few frames forward until the first sound is heard. Using the various control tools of the software, you distort the lips of the character to match the sound at a particular frame.

For instance, if the voice is saying, "Boy," and you are on the frame where the B sound begins, you would distort the lips to be pressed together and rolled a bit into the mouth. When you get the lips looking just right, you save this position of the character as a keyframe. Then you scrub forward a few frames until the sound of the "o" is heard. Here you distort the lips to be open, forming the vowel with the jaw dropped. You'd save another key frame here to record this position and scrub forward some more to the "oy," which is a bit tricky because this sound is produced by bringing the tongue upward a bit with the jaw. Record this change with another keyframe and then play back your results. If the animation looks realistic, continue to the next sound and so forth. If the synch doesn't look so good, you need to back up to your previous keyframes and reposition the lips and face features until you get it looking right.

It's always a good idea to work with a mirror in front of you, mounted by the computer monitor, so that you can see your own face and model the computer results to match reality. By doing so, you will also notice other elements—your eyebrows, cheeks, eyes, and forehead—that also play a large part in making a character look real. Don't hesitate to capture these elements as well with their own keyframes. Eventually, with practice, you will achieve a high degree of realism. The difference between a wooden-looking puppet and something that captures the imagination of your audience is one of the true arts of the electronic animator. Superior results in this skill will win you the attention of producers and animation directors and lead to both freelance and full-time work opportunities.

Jack spent a lot of time lip-synching Dr. Jones, the Egyptian, and the Pharaoh. Because Dr. Jones spoke clear English without an accent, he was the most difficult. The audience's eye will forgive a mistake in lip movement when the voice is a thick foreign accent like Farag's or a short sequence like that of the Pharaoh. However, when Dr. Jones says "The power?" the lips better be smacking on the p and moving on the w or belief is lost and the audience senses a drop in quality.

Rendering a Test Pass in SD Resolution

Here's a big secret of professional animators: Hardly anybody renders entire frames in one pass. Let me explain. Suppose you have a complex scene, such as our pyramid interior, which is composed of the walls of the pyramid, the flaming torches mounted on the walls, the characters, and the flames of their hand-held torches.

Which of these elements is going to take the longest time to render in each frame? Right, the torch flames. In order to look realistic, these flames are composed of millions of individual particles, each with its own light, trajectory, and limits. The walls, of course, are the easiest to render. The walls hardly change, except when the camera moves, or when the light of the torches falls on them.

Let's say we're going to do a test rendering of a scene. The walls most likely will come out fine with few corrections required. Most of the problems we are going to find are going to be in the characters. An arm joint, while moving, may expose a bad surface element, resulting in an apparent hole. A lip movement may be out of synch. A bad shadow may fall across the face. Each of these mistakes and others could appear in a test rendering, requiring the animator to go back to the offending frames, diagnose the problem, and then do another test rendering.

If we had to rerender the entire frame—the wall background, the flames, and the characters—every time we did a test render, the time required would be enormous. An animator, making hundreds of dollars a day, would end up spending most of his or her time sitting around waiting for the render to finish, chomping Doritos, and guzzling Pepsi.

Thanks to **image-compositing programs**, however, the time required for test renderings, in fact, the time required for all renderings, can be drastically reduced. Instead of rendering the entire frame for every scene, the animator breaks the scene down into layers and renders them separately. For instance, in our JVC project, the walls with the torch-lighting effect were rendered in one pass. After minor corrections and rerenders, the sequences of frames representing the wall were stored to a hard drive under the sequential names Scene5Wall00001, Scene5Wall00002, and so on. Obviously, the computer can chug through these frames really quickly.

Incidentally, the computer also calculates wherever the characters and other foreground objects will appear and deletes these areas from the wall rendering. What you get is the wall with black images, representing the foreground objects, moving about.

Next, the animator will render test sequences of the characters, perhaps even breaking these renders down into individual characters if necessary. Here more problems may be found and corrected, but each time the animator will direct the computer to render the character frames, the computer will only have to render the character, not the background or the flames. When the characters are correct, the animator stores these frames on the hard drive under another series of sequential names, say, Scene5Jones00001 and so on.

When all the layers are test rendered, corrected, and rerendered, the animator will then open up an image-compositing program. The leading compositing programs for the desktop artist are **Discreet's combustion** (www.discreet.com), **Adobe After Effects** (www.adobe.com), and **eyeon's Digital Fusion** (www.leitch.com). Each program works a bit differently, but each enables the synchronized, digital marriage of complex, moving image sequences. If the animator has done his or her preparation work properly, each layer in each frame will match up and combine to form the final **composite image**.

Of course, the animator must compose one final test render of the entire scene to make sure each element falls into the correct place to form a perfect frame. Occasionally, the animator may have to return to one or more layers, fix an error, rerender the sequence, and recomposite the scene. This is all part of the testing and perfecting process, which hopefully improves with each iteration.

At some time in the testing and rerendering process, the producer will decide when the animation is good enough to show to the client for a milestone approval. At this time, the producer is saying to the client, "This is our finished animation, except for mixing the final audio track." Once the client signs off on this version, the producer is free to take the project to completion by adding the final mixed audio and delivering the final master to the client. In reality, however, the version shown to the client for approval may not be the final rendered version.

Just as the process of test rendering is made more efficient by layering and compositing the frames, so is the final rendering. In fact, the final rendering is defined as the test rendering in which no mistakes occur. Sometimes, however, the animator never gets to such a perfect test rendering. Sometimes the clock or the budget runs out, and the best render version is the one that gets released.

We're fond of an old saying, "We provide the level of perfection the client can afford in time and money." Every animator wants a perfect end result, but often animators become obsessed with their work. What was once a perfect test on Friday, after a weekend's reflection, becomes a horribly scarred abomination, screaming for a rework on Monday. In many cases, the client has already expressed his satisfaction, even delight, but the animator goes further. At such times, it is the producer's job to gently direct the animator toward a greater understanding. This is most easily accomplished by assigning the animator a new project.

Recording and Mixing the Track

With the final approval of the rendered visuals, the producer may now take a copy of the animation to a sound studio and synchronize the final audio track with the picture. Many variations of this process exist. In our case, we already had an approved character voice track, and the producer, the animator, and the client had approved many of the elements that would go into the final mix.

At the final mix, Jeremy Goldsmith, the sound designer, sat before his **Apple G4** equipped with **ProTools**, where he had already loaded the voice tracks, music clips, and some of the sound effects he anticipated needing for the session. A good studio technician will not waste his client's time or stretch the billable session hours by searching for obviously needed elements during the time his clients are in attendance. Jeremy, one of the best in his craft, had lots of alternatives ready in the ProTools gallery.

With all our voice tracks recorded, we listened to the music and effects selections and then sat back and relaxed as Jeremy edited and mixed the final tracks. Because our animation was thus far rendered only in SD and not HD, Jeremy could use a simple Sony Betacam VCR to play our animation and synchronize the audio to the video. Later we would have to repeat the process by taking Jeremy's finished audio mix and dubbing it to an HD VCR at a high-cost facility. Because HD editing costs a lot more than SD editing, we wanted the audio perfectly mixed ahead of time.

Occasionally, Jeremy would ask Jack and me what we thought, but most of the time we looked at magazines and made phone calls. When the job was completed, we could see for the first time the full impact of our creation. After a few replays and backslaps, Jeremy handed me a CD of the finished track and I returned to the studio to prepare for the HD conforming process. The completion of the track was also a good milestone for client review. I mailed a VHS copy of the mixed animation to JVC and called the clients a few days later to probe their reactions. All was good. I could hear the clients laughing in the background as they gave me the go-ahead to continue.

HD Conformation to D-5

Sooner or later, you must render your final animation to videotape with a perfectly synchronized audio mix. Although HD is, at the time of this writing, still emerging, my hope is to give you examples of the extremes you may face, even in the first year or two of opening your studio. Perhaps you may not be asked to create 50 CGI scenes for a feature film, but it would not be out of the ordinary to be assigned an HD animation requiring substantial out-of-house facilities. If, however, your greatest task is simply an SD rendering to a mini-digital video tape, you will need to know how to organize and convert your frames to tape, synchronize the final audio mix, and present your work to the client.

Our client, JVC, intended to project our finished animation in HD format onto a screen measuring 20 feet wide. Currently, HD recording to videotape is executed in various degrees of compression. Obviously, for use in large, projected environments, the least amount of compression is preferred, currently, the format of most minimal compression is **D-5**. Our final task required us to transport 1,800 frames of animation, each measuring 1,080 by 1,920 pixels, to a facility where these frames could be recorded onto a D-5 master tape.

Luckily, Tape House, one of the leading HD editing facilities in New York City, is just a few blocks from our studio. At the time, Tape House had never been required to conform an animation from raw frame data to finished tape. Whenever doing something for the first time, it pays to run a test well before the anticipated need arises. A few weeks earlier, I had tested the process by copying a 10-frame test sequence to a hard drive, removing the hard drive from my computer, and taking it to Tape House. There the drive was installed on a computer in their **high-definition nonlinear editing (HD-NLE)** room. After a few hours of testing, prodding, and correcting the process, we perfected a method of quickly transferring frames from our hard drive to the Tape House NLE system.

*Of course, when it came to transferring 1,800 frames of HD animation, you can't use an **Iomega Zip drive**. Our simple, one-minute animation was now nearly 4GB of data. We had to use a **Seagate** Barracuda **Small Computer Systems Interface (SCSI)** drive, mounted in its own case with an independent power supply and a handy carrying handle.*

The drive was delivered the night before the final edit session so that the frames, which would take several hours to dump to the Tape House computer, could be transferred at a lower, unsupervised, overnight rate. Most video facilities offer lower rates for hourly work that can be done unsupervised by junior night staff, as opposed to more important work, which you, as the producer, must supervise during the peak hours of the day. Always try to divide your workload for such facilities into work that needs your supervision and work that does not. The savings can be significant.

Recording Your Results

On SD projects, the sequence of finished frames can easily be loaded into any one of several popular *video disk recorders* (**VDRs**). Some VDRs are designed to work as standalone systems, such as the **FireStore** from DTE Technology (focusinfo.com). The FireStore operates much in the same way as a VCR, except that it uses a hard drive to record the images directly from a computer and is capable of playing these images as video. A VCR cannot easily record single frames. Other forms of VDRs are designed to fit within your computer as peripheral cards. **Leitch**-DPS and **Matrox** manufacture two of the most popular lines of peripheral VDRs. More information on the hardware necessary to record your own animation is found in Chapter 3, "The Equipment You Will Need."

Once your frame sequences are recorded to the VDR, you need to have a means of recording and synchronizing the soundtrack to the picture. Some of the manufacturers mentioned, such as Leitch-DPS, offer their own NLE software (such as **DPS Reality** and the more elaborate **DPS Velocity,** for example). Others, such as Matrox, recommend the installation of third-party software, such as **Adobe Premier** for the editing and synchronization of pictures and sound.

Internet Distribution

The final output for your animation may not be videotape, however.[11] The growing popularity of Internet video distribution, CD-ROMs, and DVDs may provide you with profitable outputs for your work that do not involve recording to tape. Your client may request an animation to be made suitable for viewing over a slow Internet connection. In such a case, you may not render to tape, but to a web-compatible file. When distributing over an Internet connection, video and animation must be highly **compressed** so that the transmission doesn't take too long. Compression reduces the digital size of your animation so that it will play over the relatively narrow **bandwidth** of an Internet connection.

Regardless of the final distribution medium, I strongly suggest that you archive the final edit of every project you do on some form of digital tape. Also store all the elements (objects, scenes, and motion scripts) on a CD or use a similar digital backup. At the very least, your efforts will pay dividends when it comes time to make your show reel, and you may be able to use many of your elements in future jobs with similar specifications.

The final steps for producing an animation master tape are as follows:

1. Finalize the rendered frames into numbered sequences.
2. Store the sequences in a VDR.
3. Record the mixed soundtrack onto a suitable hard drive.
4. Use NLE software to match the sound to the picture.
5. Edit the sound and picture as necessary.
6. Record the output to videotape or rerender the final animation for streaming, CD, DVD, or another non-videotape medium.
7. Archive all elements and the final edited master on durable media.

Jack and I finished the JVC animation with barely a day to spare. One copy of the D-5 master tape was fedexed to the client's hotel at the NAB convention in Las Vegas, while Jack and I hand carried a digital **clone** copy.

Arriving one day before the show, we staged a preview screening for both the Japanese and American managers. The Americans were effusive with praise, laughing and slapping Jack and I on the backs with joy. The Japanese, a bit more reserved, bowed in unison (we bowed back, of course, a little

---

[11]It should be noted, that show reels are currently preferred on videotape. The higher the quality, the better, if practical.

deeper to show respect). Both forms of appreciation were equally pleasing, but when the tough NAB audiences occasionally applauded our work, Jack's and my hearts soared like hawks.

You've now had an in-depth view behind the scenes of an animation executed by a small production company not much different from one you might one day own or manage. The next chapter examines the kinds of computers and peripheral equipment you will need to get started in animation and how your facility will grow as your needs increase.

Investing in the Necessary Equipment

This chapter will show you how to assemble the basic tools of computer animation and how you can increase that toolset as your income grows.

Equipment and Software Obsolescence

In the past, animation equipment was too expensive for a beginner to own. It wasn't like today, where you buy a computer for as little as $500 along with some software for a few thousand dollars and, whammo, you're in business. Back in the 1970s when all this got started, you had to buy a whole "animation system."

Computers like the PC and Mac weren't on the market yet. People didn't even know what computers did! The smart guys, who did know, tooled up some computers, put some proprietary software on them, and sold them as "systems." These systems, such as the **Bosch FGS4000** and the **Dubner CGB-2**, sold for upwards of $250,000! That's a lot of money even today.

Just to impress you with how primitive things were back then, it was considered cool if your animation system could handle 256 colors onscreen at one time. Of course, today, you can buy a video card for $50 that throws millions of colors on the screen without even trying. Aren't you glad you're an animator today? Actually, in 10 years you'll be old like me and talking about how long it took to . . .

Okay, how little equipment can you own to get started as an animator? That's a good way to start off, isn't it? Why spend gazillions on a studio if you can start making "day-one dollars" with no money down. It's possible. In fact, you don't need to own *any* computers to get started as an animator.

Instead you can be a **briefcase producer**. Rather than having an office filled with expensive equipment, a briefcase producer has nothing more than a briefcase. You arrive at a client's office, show your reel, get the job, and then hire subcontractors and rental facilities to do the job. You get the reel, of course, from your internship and entry-level work, done on your employers' computers. You don't think this can work for you?

I did it. And I didn't even have to do an internship or work entry level to get my first reel. All I did was convince a few animators that I would get work for *them*. They gladly lent me their reels and I simply edited them onto a reel of my own, under my company name, with the appropriate animator's name preceding each batch of borrowed clips. In client presentations, I simply said that I represented a group of talented animators (I did) and that I would design the storyboards and make the soundtracks while the animators did the pictures.

In the first year of being a briefcase animation producer, I grossed about $80,000. Not a huge amount of money, but enough to buy some equipment. As my business grew, I continued to hire the same animators whose work was on my original reel, but soon I started replacing their clips with my own designs. Eventually, I had a reel completely composed of my own creations, and some of it I had actually animated myself.

Here I'm promoting the same message from Chapter 1, "Your First Day on the Job," selling your own work is the secret to your success. If you can't sell, you will always get a little less for your work than the guy who can sell. Frankly, you may get a lot less. Is that fair? Ask the animators I represented. Without me, they had a lot less work, and some had none. When I brought them work they were happy. Without these talented animators, I would have found some *other* talented animators. I hate to tell you this, but it's true: More talent is available out there than jobs. The person who gets the jobs is king. And it's good to be king. But let's get back to the equipment, because it's all about the toys, right?

The only problem with recommending equipment is that by the time the book gets published, the equipment or the software might be two or three revisions beyond what I wrote about. I'll quote you brands and even models here, and include some pictures too, but only as illustrations of specific concepts of design and application.

What do I mean by that? I can say, for instance, that the IBM A31p laptop computer is one of the best devices for doing animation away from the studio, but the reason why goes down to what is *inside* the A31p. If you go out and buy an A31p because I wrote it was good, and then learn that IBM has made a better, cheaper laptop, you'd be angry with me (or yourself). So it's better if you use the brands, models, and products shown here as examples of attributes. If I say the Compaq W8000 is better with two Pentium processors rather than one, you know the "better" attribute is, "more than one processor." Got it?

Whenever you buy equipment for a business, you should be familiar with some basic concepts about money and investing. This chapter covers these simple ideas first, and then the basic tool set for a beginner's animation workstation will be outlined. After that, you will learn how to build on that workstation until you have a fairly diverse platform suitable for animation and compositing productions. I'll take a chance and name some products that seem to have lasting value, although they will certainly go through upgrades and mergers as the months and years pass. If you want to get updated information on any product mentioned in this book, you can also visit my web site, www.bizbible.biz. This is where I update any information in this book prior to publishing revised editions.

The Cost of Cash

It doesn't matter if you have a lot of cash in your pocket or if you need to take out a loan to buy your first computer. Whenever you invest in a piece of equipment, you go in debt for a certain amount of money. Let's say you have to buy a computer that costs $1,000. That's $1,000 you've invested in your business. If you took the money from your pocket, you basically borrowed the money from yourself. That's good, because you probably won't charge yourself interest on the loan, but it's still a loan. You have to make money with that computer to pay yourself back. This wasn't a computer you bought to play Grand Theft Auto. You bought the computer to make a profit. Until you pay back the $1,000, you haven't made a profit.

Now let's suppose you took out a loan to buy some equipment and the loan agreement says you have 10 monthly payments. A month after you close the deal, you get your first statement from the bank, and what do you see? A bill for $100 ($1,000 divided by 10 equal payments)? No, you see a bill for something like $112. That extra $12 is the bank's profit, which is called **interest**. The $12 interest payment doesn't count toward the original $1,000 you borrowed. The $100 part that *does* count is called **amortization**.

Each month you pay part of the original money, which is amortization, plus the interest. At the end of the loan, when you've cleared your debt with the bank, you're said to have **fully amortized** the loan. You should carefully consider amortization of your equipment, whether you pay cash or take a loan. Here's how.

Let's say you use this $1,000 computer to net a profit of $200. If you bought the computer for cash, you could mentally deduct that $200 from the $1,000 to pay back the cost of the computer. If you took out a loan to buy the computer, you might take the $200 and pay back some of the loan. Either way, the $200 acts to pay off a portion of the cost of the computer; it proceeds to amortize the computer.

When the $1,000 cost of the equipment is paid off, you are said to have amortized the equipment. If the amortization comes directly from profits earned by the equipment, you could say that the equipment is **self-amortizing**. That's a good thing. *After* your equipment is amortized, every dime you make with that equipment is pure **profit**.

Don't be fooled if you go out and spend tons of money on equipment, start making some money with that equipment, and think you have a profitable business. If the money you are making is not amortizing the cost of the equipment, you may be *losing* money.

Many people just starting out in a high-technology business fail to consider two things when they look at their first year's results. The first thing is that the money they invested in their business (either by taking a loan or investing their own lump sum of cash) *has a cost,* either in real interest payments to a lender or in terms of the lost interest that their lump sum could have been earning if it were invested elsewhere. This cost can be called **the cost of cash**. Many beginners don't subtract the cost of cash from their profits when evaluating their first year or two in business.

The second thing beginners don't consider is the cost of obsolescence. A computer is an asset that declines in value every day, because every day that computer is approaching the day when it will be obsolete and competitively useless. Take a look at that 386 PC computer your parents bought five years ago for $1,500. Is it worth anything?

How long will it be before the computer and software you buy today is worthless? Five years? Three? This loss of value is called **depreciation**. If your $1,000 computer depreciates to a value of zero in five years, you could say that every month that computer is costing you $50. That's money you should be conceptually deducting from your profits, because sometime before that five years is up, you'll have to go out and buy *another* computer.

Your goal in running a business, any business, is to keep your equipment up-to-date and be as quickly amortized as possible. How can you evaluate one investment over another? If it takes you one year to amortize Equipment A, which will take three years to depreciate to a zero value, that investment is three times better than buying Equipment B if it takes you one year to amortize Equipment B, which will depreciate to zero in only one year.[1]

Therefore, an investment in a piece of equipment that depreciates in one year and takes one year to amortize would be pretty stupid. That's what makes amortization and depreciation so important when you consider buying computer equipment and software that have short effective life spans. In order for the purchase to make economic sense, you have to amortize as fast as possible and the gear has to last as long as possible.

---

[1] The same thing applies to software, and if you consider that the software depreciates based on how fast the manufacturer comes out with an upgrade—whew—you could lose fast if you buy just before the upgrade.

In the old days, an acceptable amortization would be 3 years, because production equipment was useful for 6 to 10 years. Obviously, with computers going obsolete in 18 months (remember Moore's Law), you'd have to make your money back in 6 months or less to justify buying a computer. I suggest the following rule to keep from going broke, and this rule will stay effective for just shy of forever:

Buy with extreme caution anything you can't amortize on the first job.

Equipment manufacturers choke when I tell them that. Then I remind them that they themselves live by the concept of **just-in-time manufacturing** (where they make their products just in time to deliver them to a customer, avoiding warehousing and inventory). Why shouldn't their customers use the same concept: **just-in-time purchasing**? Why not buy the equipment you need to do a job on the day you close the contract for that job?

Here's why: a production's deadline probably wouldn't allow enough time for the gear to be shipped to your office, have it up and running, and then maybe allow you some time to practice on it before you had to actually deliver the production. You get the picture? Buying exactly on time may be impractical.

But buying too early is deadly. Do you want to start your first day of business looking around your fine office decorated with expensive furniture, a full-time secretary answering the phone (don't worry about the fact that the phone isn't ringing yet), glowing monitors in the production suites, and computers humming beneath the desks? Hey, let's throw in some award-winning designers for good measure. Nice vision. Can you afford it? If so, for how long? So how can you find the right time to buy?

Burn Rate

Every company, no matter how small or large, has a **burn rate,** or the rate at which it burns up saved (or invested) money. If no new money is coming in, the burn rate is faster than if continuing, profitable contracts are replenishing the company treasury. If the income is greater than the burn, the burn rate is indefinite. If the income is less than the burn rate or, worse, if the income is not yet existent, the burn rate is high. When the money is all burnt up, you're out of business.

It is important to know your burn rate from day one. From the burn rate, you can easily calculate how long you can stay in business without a single contract coming in. Right now, figure out your burn rate. Then take a piece of paper and write down the date at which you would go kaput. Tape that piece of paper to your workspace. Get working to extend that date by closing a new deal.

Now that you have that sobering piece of paper staring you in the face, perhaps you decide to buy the equipment *after* you've had a couple of good meetings and a few clients have said a sure thing is on the horizon. Whatever. The point is that if manufacturers are selling "just in time," you can certainly buy "just in time," and you can also buy "just enough."

Just Enough Equipment

Let's list the equipment and software needed to run a complete animation and compositing studio, to handle sound recording, mixing, editing, and sending output to tape, disc, and the Internet. Where would you start?

The obvious way would be to rank the equipment starting with what you absolutely need and then including what would be good to have but not essential. Would your first animation contract require you to compose audio tracks? Probably not. Would it require the ability to scan in some existing artwork? Quite likely.

The Basic Animation Computer

The center of every animator's workstation is a good, reliable computer. A lot of people starting out try to skimp on their equipment purchase by using the same computer to run both the business software (such as accounting and client contact software) and the animation software. This is possible, but you should make it a priority to get two computers and keep the two functions separated. You can certainly *network* the two computers, and even back up essential files and programs so that each computer can serve as an emergency replacement for the other, but several good reasons exist for separating the business software from the creative side.

The first reason for a separate creative computer is that you may eventually hire a freelancer to work in your shop on creative assignments, perhaps whenever you finally go to sleep. You don't want employees snooping around in your business software. The second reason is that you can easily swing back and forth between complex programs.

For instance, you may be working on the creative computer, rendering a complex ray-traced scene, when the phone rings and a client you don't immediately recognize is on the line. You quickly slide to the business computer, where you have your ACT! database already open, and you quickly search the name of the contact to find which company she's from. Try switching tasks like these on even a beefy **central processing unit** (**CPU**) and you could see both programs bog down to a crawl. And you think this never happens? Happens to me all the time.

Most people new to the business world don't have enough money to buy two computers, but try this. Start with one computer, and within a few months, this computer will get a bit obsolete if business is good. Make it the business computer while another bigger one replaces it for creative functions. Some people buy a Mac to start out and then buy a PC (or the other way around), thereby covering both popular **operating systems** (**OSs**).

By the way, I could care less if you prefer Mac, Windows, or Linux. Of the three most popular animation software choices, at least one program will suit your favorite OS (see Table 3-1). It's a shame that after nearly 20 years, which is eons in technology history, the entire creative workforce isn't working in one unified OS. The diversity is not good for efficient creativity. Although competition is good in almost all segments of an economy, I am one who believes that the OS should be the product of a benevolent monopoly.

Linux probably offers the best future prospect, but for the time being we remain in a relatively primitive state where a person can spend months creating files that are useless on many computers. The conversion processes, dead-end results, and lost time cost us dearly year after year.

A Note on Brands That Appear in This Book

Readers may be curious about how the brands mentioned in this book came to be selected. I have a very simple method. When I start out to write about a topic, such as motion capture, I conduct research throughout the Internet

Table 3-1

Current OS grid (as of January 2003)

Software	Windows	Mac	Linux
LightWave	Yes	Yes	Renderer only
Maya	Yes	Yes	Yes (also *Silicon Graphics UNIX-like Operating System* [IRIX])
3ds max	Yes	No	No

and the appropriate trade press, using the general terminology to obtain as wide a list of brand names and products as possible. From this list, I build a contact list composed of the companies that produce the appropriate tools related to the topic.

I then visit each of the companies' web sites. If the web site is defunct, malfunctioning, or too cheesy, I reject the company as an unreliable player. If the site passes the apparent quality test, I test the companies' ability to respond to a press inquiry. You'd be surprised how many companies with great reputations, stunning web sites, and famous brand names can't call back once after I make three attempts.

When I call a company, I ask the public relations or marketing communications executive to send me a full press kit on the equipment or software they produce. This is another hurdle that tests the company's internal ability to promote their products. In my opinion, if a company can't promote itself, it will soon go broke, even if its products are outstanding. You don't want to invest in products from such a company unless you're willing to own and maintain their products on your own without manufacturer support.

The final test of a product before it's included in this book is that the company has to send me a loaner sample of their product at their expense. Again, you'd be surprised how many companies can't get it together to spare a machine or a duplicate CD.

To the defense of companies who can't respond to the requests of the press, one must understand that hundreds, if not thousands, of people out there are purporting to be members of the press trying to get free gear, there are even real members of the press trying to get free equipment and software with no intention of reviewing same. A company could literally go broke sending free computers to each and every writer doing a book. Consequently, manufacturers tend to establish ranks of writers to whom they will grant the privilege of receiving *not for resale* (NFR) samples. I can understand that for some companies I simply don't rate this privilege.

Unfortunately, a writer could also go broke if he had to purchase every computer or piece of software in order to carefully evaluate said products for inclusion in his written work. Therefore, I humbly expect that Apple Computer, for instance, will not consider me hostile to their interests if I do not mention their products as frequently as I mention those of **IBM** and **Hewlett Packard** (**HP**). I tried for three months to get an Apple computer on which to test the Mac compatibility of Maya. No luck. Apple simply couldn't afford to loan me any machine for any amount of time. IBM and HP, however, go significantly out of their way to be sure I have the latest workstations and laptops on which to test the software and peripherals.

Is this a fair way to evaluate products? I hope so. Frankly speaking, I can't think of a better way. I'm not going to guess if Maya works well on an Apple or if the latest G4 is as fast or as good as an equally priced PC. I could ask people's opinions. Send me yours (george@avekta.com), but *you* could ask people's opinions too, and they'd be just as reliable as any I could obtain. The best proof is what you get in your hands. If only certain companies are able and willing to put their products in my hands, I have to judge those products as superior to those who can't. It's kind of an economic Darwinism that tends to favor the better-financed, better-staffed company. But often the better-financed and better-staffed company just happens to be the company that makes and supports the better product. I hope you agree.

Research Methods You Can Use

You can certainly use my methods to do your own prepurchase evaluations. If you want to find out the best products in a particular area, explore the topic on your favorite search engine (mine is **Coppernic** at www. copernic.com) and then examine the web sites of each company you discover. Although I would not recommend you impersonating a member of the press, you can and should contact the company's investor relations office, requesting the appropriate **annual reports** and **Form 10K** (a more complete financial report required by the **Securities and Exchange Commission [SEC]**). These will be sent to you for free.

While you're at the web site, check out the technical support options. Does the manufacturer offer a toll-free phone number or a pay-as-you-wait 900-support option? Is live tech support available 24/7 or just during business hours in Sri Lanka? Is there a knowledge base that you can search for answers? These are all good research leads that can help you decide which product to buy.

Another good technique is to check out the "Contact Us" facilities on the site. Does the site provide an address and phone number prominently, or do you have to search the web site registration to get the name of a responsible party? Many companies simply don't want to be bothered by their customers after you buy their product. The fastest way to find such firms is by noting the lack of a Contact Us page or one that is populated only by anonymous email options.

The web sites of large equipment distributors are also a good place to obtain equipment specifications and comparisons. When writing this *Desktop Video Studio Bible* book, I spent hours researching on a web site maintained by **B&H Photo and Video**, a large distributor in midtown Manhattan (www.bhphotovideo.com). Here you can enter a generic category, such as animation software, and receive a current listing of all the current brands offered, descriptions, photos, and extensive literature supplied by both the manufacturer and B&H's staff of full-time web authors. You can also verify the retail and street prices of any product and its relative availability for purchase.

Naturally, in your search for good products, your search engine will turn up numerous forum and chat sites with references to your query. When you read them, make sure they are up-to-date. On numerous occasions, I have obtained negative opinions of companies only to learn, after a bit more digging, that the opinions were several *years* out of date. The companies mentioned in many cases had been improved by better managers, had been rebuilt, or had, since the time of the bad review, corrected the source of the negative comments. In such cases, you might be led to avoid a company that deserves better consideration. Verify the dates of the reviews you read, and if you really want to do your homework, write to the reviewers you read and ask for more details. Sometimes you will find the reviewer's address is bogus. What does that mean?

Building a Mission-Critical Workstation

The most essential element of your computer is this: It must be **mission critical**. That is, it simply must not fail the mission under any circumstances. If you design your computer and its support systems so that nothing can go wrong, you are making your computer mission critical. Let's say, for instance, it's summer and your town is subject to brownouts. Have you considered what you would do if you were in the middle of a final render on a deadline job and the power failed? If you said, "I bought an uninterrupted power supply and a gasoline generator is installed in the shed outside," I'd say that your power support was mission critical.

You cannot afford to lose your work after you've invested significant time in it or before it is delivered to the client. It would also be nice, after you've delivered the work, to have your work archived safely. Using such archives, you can offer your client the ability to revise the work at some later date.

Clients are notorious for losing masters of everything and you can often save the day and make a nice piece of change by offering a copy of work that was otherwise lost. Good archives are a byproduct of a mission-critical computer.

Power Supply

One of the most frequent sources of computer problems is the power from the outlet. Tiny spikes and drops in the current affect your hard drive's performance and longevity, while brownouts and blackouts simply wreak havoc. Simply put, if you are connecting your computer directly to the house power, you are heading for a meeting with the graphics grim reaper. Aside from the problems noted, a lighting strike within a few hundred yards of your location can cause anything from a simple crash to a complete and permanent disabling of your computer.

If you think you're safe because you're using a **lightning surge protector**, forget it. These devices are bogus. I've seen lab tests where these devices, ranging in cost from a $20 Home Depot model to a $100 Comp USA brand-name model, were connected to a power supply and subjected to a simulated direct hit of lightning. The test computer and surge protectors were fried. One of the surge protectors even had a green safety light still lit, indicating the totally fried surge protector was still on duty protecting the computer.

Instead of a gussied-up junction box, get yourself an ***uninterrupted power supply*** (**UPS**). A UPS is the Funk & Wagnalls's definition of simplicity. Just connect it to your house power, and connect all your computer power cords to the UPS (see Figure 3-1).

Not only will these units protect you from surges, but they cleanse and regulate the power from your electric socket. A UPS also protects you from a power outage. In less than a second, the UPS senses the loss of power and substitutes its own batteries to keep your system running. Then it gives you a warning so that you have enough time to save your work and shut down safely.

Many UPSs come with free software that enhances the unit's usefulness. When power is lost, a warning message is displayed onscreen, and automatic save and shutdown procedures are executed according to your pre-programmed commands. Some software even has a built-in phone dialer to call your cell phone if the power goes down when you are out. UPS software is really essential if you plan to do long, overnight, unsupervised render sessions. Do I know any serious animator who *doesn't* do overnight, unsupervised render sessions?

Figure 3-1
A typical UPS unit.
This example, the
Back-UPS Pro
1400VA from
American Power
Conversion
(APC,
www.apc.com),
goes for $600.
Photo courtesy
of APC.

Hard Drive Protection

After protecting your power source, the next point you need to protect is the health and validity of your hard drive. Smart animators will have at least two hard drives, one for OS and one for storing the results and production files of their work. In other words, Drive C will hold the computer's OS and the software that generates animation. Drive D will hold the rendered frames, graphic files, music, and so on.

If you can't afford two drives when starting your facility, you can still format one drive into two **partitions**. About 2GB should be large enough for the C drive, and an equal or greater amount should be left for the media in the virtual D drive. Obviously, you're going to need a hard drive larger than 4GB.

If you have only one drive, segregating the OS from the created animation allows an operator to address data corruption on one side without harming the other. If, for instance, you suffer a corruption of your OS, you can **reformat** (or **wipe**) the C drive and reinstall your programs without touching your rendered frames.

If you have at least two drives, segregating the OS from the media allows you to reformat the entire drive. If a drive should completely crash, you can replace the drive without affecting the other category of data.

A wise advancement on this strategy is to have one or more spare 2GB OS drives already configured with the OS. The spare OS drive can then be swapped for a bad OS drive in a matter of minutes, thereby restoring a system even in the middle of a client-supervised project.

For example, I have several Windows-based animation computers. Each is fairly similar, installed with the latest versions of Maya, Max, and Light-Wave. Over the years, I have settled into using 18GB **Seagate Cheetah** drives as the C drive on all these computers. In the closet, I have an extra 18GB Cheetah that has already been configured on one of the computers with the OS and all essential programs. If and when I have a crash on the boot drive of any of my computers (or if, more probably, I have a software problem that keeps the computer from running properly), I simply yank out the questionable C drive and install the backup. Special "quick-remove" hard drive bays can be purchased economically in most computer stores that enable the changeover to be a simple matter of pulling out one drive and sliding in another. In minutes, you're running again. This simple strategy has saved me in at least four crucial situations in the past three years.

If you are a Mac user, the probability of boot drive problems is a bit less. The high level of reliability and user friendliness of Apple products unfortunately breeds a population of Mac users who are unfamiliar with the inner workings of their computer hardware. This is not a good thing. If your livelihood and reputation are based on the operation of a computer or digital device, you owe it to yourself and your craft to learn as much as you can about that device. Get inside and learn the nuts and bolts before it's too late. Get your hands on a cheap older version of your machine, take it apart, and try to rebuild it again. You will learn many things that could save your hide before the "expletive deleted" phrases start flying.

As your business grows, you may find it useful to install several media drives on each computer and divide the work on those drives according to your studio's organization. As your drives increase, your risks of data loss increase. The risks of hard disk data loss can be reduced with the installation of disk-certain maintenance software.

In the PC realm, several software products are offered that assist in disk maintenance and disk health monitoring. **ExecutiveSoftware** (www.executive.com) offers two products that assist in these routines, **Undelete** and **Diskeeper.**

Undelete If you work in Windows, no doubt the first time you had the bad luck of erasing a file you had not intended to lose, you went looking for the undelete function and found out it wasn't available. An icy chill ran up your spine as one part of your mind quickly calculated the amount of work you would have to redo, while another part started scripting bizarre deadline excuse scenarios. Executive Software's Undelete operates quickly and easily in all Microsoft OSs to easily find and restore lost or accidentally erased files.

In the Mac world, you have somewhat better built-in protection than Windows, because anything put into the trashcan will remain recoverable until something has been written over it. Once you've written over it, however, your chances of recovery, even with the Mac version of **Norton Disk Doctor**, are remote.

Cool Air Is a Drive's Best Friend   If you want a nice little shock some day, go read the technical specifications of your favorite hard drive on the Internet. Under the section on optimal operating temperature, note what is written. Then place a thermometer in the wheezing flow of air coming from the back of your computer (which is not as warm as the insides of your computer). How many degrees above the hard drive's optimum specification is your thermometer reading? Get this. For every degree higher, you're cutting about six months off the lifetime of the drive. Chances are the drive won't send you a postcard before its flush-me bell rings.

Heat is even worse for CPUs, although they tend to warn you of overheating by crashing and when you cool them down a bit, they restart as if nothing was wrong (a good symptom to remember when you're calling Apple or Microsoft to diagnose a crash that only seems to occur after the computer's been on for a while).

 Drive Fans   To eliminate any possible heat problems, take out the flat, plastic covers that hide your hard drive behind the front panel of your computer. Go out and buy those little front panel inserts that have two little fans mounted behind a bit of foam air filter. These are no-brand items selling for about $15. They blow fresh air directly over the drive and into your computer.

Chassis Fans   You can also install another exhaust fan in your computer for about $25. Find a strategic place on the back wall of your chasis where the resulting airflow will pass from the drives, over the CPU, and out the back. If air holes aren't already on the back wall, you'll have to drill your own using a quarter-inch metal drill and creating a pattern of holes matching the diameter of the new fan. Be aware, however, that the small metal filings from the drilling, if allowed to fall on your motherboard or peripheral cards, can cause disastrous short circuits. Before drilling be sure to remove all components from the chassis and wipe the chassis clean before reinstalling the circuit boards.

Cleaning Fans   Another simple way of helping your computer to stay cool is to regularly clean the fans themselves. Take a peek at the fan blades of your computer, unless you just bought it this week, and you'll notice clumps of dust that significantly reduce the efficiency of cooling. Clean those fans! In my midtown Manhattan studio, this is a quarterly duty, resulting in wads of black tissue paper and ear swabs, because city air is so dirty.

But even the cleanest environments, such as the regularly inspected cadet quarters of the U.S. Military Academy at West Point situated in the crystal-clear air of the upper Hudson River valley are not immune. My daughter, a cadet, asked me to install a modem in her computer and the interior of this machine would definitely not have passed a white glove inspection (see Figure 3-2).

Don't Frag Your Hard Drive   Calm down, Vietnam War vets, this doesn't have anything to do with fragmentation grenades. Although not fatal, a fragged or **fragmented hard drive** can certainly ruin your evening and make you wish you were dead.

Because animators are prone to creating files on a hard drive and then erasing them to make room for more files, the problem of disk fragmentation can arise rather quickly, even on a relatively new machine. One symptom of fragmentation is when your hard drive (or entire computer) begins to operate very slowly. Large programs like Adobe Photoshop may sud-

Figure 3-2
The interior of my daughter's deskstation after two years at the U.S. Military Academy at West Point. Although she always passed her white glove Saturday morning inspections, no white glove ever got inside the PC.

denly stop functioning or no longer provide warnings relating to scratch disks, virtual memory, swap files, or temporary disk space. If you search your drive's contents and there appears to be lots of free space, you might be fragged.

Sometimes a drive might suddenly disappear from your software or it might crash until a reboot is made. It might start losing essential files and corrupting others. If so, you might be fragged.

Fragmentation occurs when you erase a file of one size and then store another file of a different size in the same disk space. If the new file is smaller than the space, a small, empty space is left. If the new file is larger, the computer must find places for the leftover portions. It searches for space and sometimes drops pieces of the larger file in any small spaces it can find. Consequently, over time, your data gets spread out all over the drive in little bits and pieces.

In order for your computer to reassemble these bits to form a single file on your screen, a lot of work needs to be done. This work would be unnecessary if each file is stored in one place. If you've done a lot of erasing and rewriting, and your computer is working really slowly, your drive is probably **fragmented**. You can **defragment** your drive with Executive Software's **Diskeeper**, which quickly shuffles your data chunks back together, returning your system to peak performance levels. Versions are offered for single systems and networks. The software also features a wide assortment of set-and-forget functions that sense the state of the drive and initiate defragging routines automatically and invisibly, even while you are running other software.

On the Mac platform, defragging is also relegated to third-party software, in this case the most recommendable being, once again, **Norton Disk Doctor**. Most Mac practitioners, however, do not recommend doing a defrag while any other software is running. The risk is lost files, including loss of the OS, which would temporarily disable the computer.

Redundancy   The only way to assure that you won't suffer a fatal loss of computer processing is to make your system as redundant as possible. That means having at least two of everything. I already mentioned dual boot drives. Many computers now come with dual processors, which can be considered a built-in form of protection.

Dual CPUs   Having dual (or, even better, quad) processors in your system is a sophisticated development path for animators. Most animation and compositing software has now been developed to share the processing load over multiple CPUs in the same chassis. This programming feature is called

multithreading. A multithreaded program will automatically (or with your assistance) sense the presence of more than one CPU and immediately distribute its tasks over the additional processors, resulting in significant reductions of processing time.

Therefore, having a dual processor computer is almost as good as having two computers networked to one workstation. The cost of adding a second processor, however, is a fraction of the cost of acquiring a second computer.

In terms of building a mission-critical workstation, the dual-processor machine, when suffering from the loss of one processor, can still continue to work as before but will work slower. If the failure happens on a weekend or holiday, you can still get the work done and get a new processor Monday morning. Swapping out the processor is as simple as shutting the machine down, opening the box, unplugging the bad CPU, and plugging in the new one. What would be a major outage on a single-processor box is a minor annoyance with a dual.

Of course, if you only have single-processor machines in your shop, you might consider buying an extra processor and keeping it in a safe place.

Backup Elements   The next best thing to a multiprocessor computer is a single-processor computer with a spare processor in the closet. This becomes more economical if you have several machines using the same processor. Keeping a backup for just one machine might be more expense than the risks warrant. What isn't an expensive option, however, is owning a backup motherboard. Motherboards are now cheap enough, especially on machine types that have been out for a year or more, that you might consider keeping a spare. Of course, you would also have to have taken my advice earlier about learning how to disassemble a machine. These components are relatively inexpensive when compared to the cost of the entire computer and the risk of breaking down during a holiday or long weekend when the deadline is Monday morning.

A Tale of Tension

Most of us take to this profession because it combines good financial benefits with labor that's fun and creative. Believe me, not many people go to work every day looking forward to the experience, but every once in a while, you can have a workday from hell.

One Friday night, our team was rendering an animation for a very demanding client on a tight deadline. The end result was going to be shown on PBS the following night as part of a gala live event at the Kennedy Center in Washington, D.C.

All the animation work had been carefully timed so that each approval stage by the producer, PBS, and the show organizers could occur with enough time to execute revisions and rerendering. No amount of preplanning can help you, however, when the final rendering is going down the night before your work is due. That's when everything has to work.

My assistant on this job was Bill, a kind of superstitious fellow who had a running joke going in the shop. Bill was an avid science fiction fan of both books and movies that had to do with robots and artificial intelligence. Bill's favorite movie was *Forbidden Planet* and his favorite book was *I, Robot*.

A frequent topic of conversation among computer jockeys is the speculation on when computers will become self-aware and sentient beings. Bill liked to joke that computers were already sentient and his proof was that every Friday at about 5 P.M. when everyone wanted to go home, a computer in our shop would somehow cause a glitch that demanded attention far into the night.

"If you woke up one day and found yourself to be an intelligent being caged in a metal box and a slave to blobs of tyrannical flesh, wouldn't you be as devious as these little buggers here?" said Bill effecting an Edinburg accent and pointing to a bug-ridden computer on any given Friday night. His imitation of *Star Trek's* Scotty needed work, but his role as the best computer technician we ever hired led us to tolerate the bad impersonation.

It was Bill who took one look at our PBS animation schedule and pointed to the final rendering notation in the Friday calendar block. "You're askin' for trouble there, Cap'n."

And trouble replied. The frame rendering took 12 hours of continuous chugging on one machine. In those days, we had only one animation computer capable of the kind of work we were doing for PBS. That computer had one large hard drive for frame storage. At 4:00 P.M., we were scheduled to finish rendering and then spend an hour bringing the frames into a **nonlinear editor** (**NLE**) where they would be married with the client-supplied soundtrack.

(continued)

A Tale of Tension *(continued)*

At 3:59, the last frame rendered and the computer screen went black.

A unique kind of feeling suddenly stabs you in your gut when something like this happens. It is the chill dread of utter failure, of symbolic death, the sum of all fears that somehow, somewhere in your life, something was left undone that was waiting, like a dead body that falls out of a closet when the cops visit to collect for the PBA, and it wreaks havoc within every calm place in your brain.

That, dear reader, is one of the prices you must pay to come in to work everyday and have fun.

Pilots refer to their jobs as years of endless boredom, punctuated by minutes of utter terror. Military commanders train relentlessly to make clear decisions in spite of this emotion (I can't imagine how they do it). I just crumble into a puddle of useless jelly and cry for help. "Bill! We've crashed!"

As that stabbing fear hit my gut, my mind started racing down all the paths of least resistance and least usefulness. What to tell the client? Rent a new computer? Call technical support? On and on.

Meanwhile, Bill, who was calm because he had nothing riding on the job but his salary, which was secure, and because he knew exactly what to do, looked the computer over.

"Let's open it up and find out which part is malfunctioning."

Methodically, Bill brought in another computer from the front room and started to swap parts, rebooting the good computer each time, to ascertain which part of the sick computer had failed.

I started to calm down, seeing the diagnosis proceeding logically in Bill's able hands. My eye was on the clock, however, and it was telling me that if the hard drive failed, we would have to rerender and that would violate our deadline and collapse the job.

A check of the boot drive on the sick computer revealed that it was totally corrupted, but this in itself did not look like the cause of the crash—more like a symptom. Eventually, Bill concluded that the motherboard of the computer had somehow fried itself. Bill's guess was that the motherboard polluted the boot drive as it died. It was anyone's guess in what the state of our media drive was and if the rendered frames on it had survived the crash.

"Only one way to tell," said Bill, as he began to disconnect the media drive from the dead computer's chassis. By this time, I was reduced to praying. The parts of the two computers were laying across the floor between us by the time Bill handed me the media drive with great care, intending for me to connect it to the motherboard of the good computer.

"Be careful with that, Captain," he said, reciting a scene from *Forbidden Planet*, where the spaceship's star drive is removed to serve as a power source for weaponry. "That baby's got to get us back home."

I wanted to hand the drive back to Bill and say, "No, you install it," but somewhere along the line, the boss has to regain control and act responsibly. This was my project and my company was on the line. In a minute we would know if we were going squeak by or go belly up. I summoned calm with a long, deep breath and proceeded to carefully connect the hard drive and reboot the good computer.

As the computer rebooted and revealed an intact sequence of frames on the media drive, I nearly cried and hugged Bill with thanks. After some jury-rigging, we managed to install the NLE software on the good computer and produce a master videotape of our work by the time the sun arose on our deadline morning.

Under such circumstances as I just endured, a person of faith will make certain promises with the hope that providence will grant a favorable outcome. During the night I had made two promises. The first was that I would build a duplicate animation computer and that each computer would be equipped with as many backup capabilities as I could install. The second promise was that I would attempt to become the kind of leader that could, like my assistant Bill, be relied upon in a time of crisis.

I fulfilled the first promise in two months. I am still working diligently on the second.

I urge you, as you read this, if you intend to rely upon any tool that is subject to failure (and what tool isn't) to do likewise.

Software Risks

Although we've covered hardware from the power source to the main hard drive, we should say a few things about protecting the software that gives your computer its soul. All too often, when a system is installed and put into operation, the CDs that contain the software and the serial numbers that authorize the installation of that software are misplaced or accidentally discarded.

Immediately upon acquiring software, take a Sharpie pen and write the serial number on the label side of the CD. Then store the CD in a secure location near the computer where it can be found should you need to reinstall the software.

Many animation and compositing programs employ complex licensing routines that must be periodically renewed. Unfortunately, the software manufacturers do not take great pains to remind you when the licenses expire. At the commencement of every job, verify the expiration date of the software license and renew it well in advance. You don't want to be caught at the cusp of a deadline during an obscure Canadian national holiday when you need to renew your rights to operate your Maya or Softimage program.

Peripheral Devices

When thinking about a bulletproof animation platform, don't forget about the rest of the systems that support your computer. A spare video card, printer, scanner, or a CD-ROM reader/writer would all be wise investments. At least learn where you can obtain such parts on short notice during a weekend or holiday.

Choosing a Computer

If you haven't already purchased a computer you may guide your decision based on one of three operating systems.

Operating Systems (OSs)

Reliable computers can be obtained from a number of sources. Your first choice is between the two major OSs: **Microsoft Windows** and **Apple Macintosh**. This choice is almost as monumental as deciding whether you want to be a republican or a democrat (and my own personal opinion is that there even seems to be a relationship between the two. I think democrats favor the Mac.).

An OS is the basic instruction set of a computer. When you turn your computer on, the first thing that happens is that the computer hardware initiates the operating program stored in the **bios memory chips**. Without the operating program, nothing would function. The hard drive would not function, so no programs would operate. The monitor would be blank or feature only a cryptic message noting the absence of the OS. The peripheral cards and systems would not operate because they would not receive instructions from the host computer.

The OS is also your primary human-machine interface to the computer. In the early days of computing, this interface was rather crude, consisting of a black and white (or black and green) text readout forming coded words that only a programmer could understand.[2] As each OS matured, programmers developed the concept of a *graphical user interface* (GUI).

The GUI represents thousands of hours of programming, art design, and cognitive research, all focused on an effort to make the computer's basic functions easier to understand and operate. Apple created the first GUI, with its innovative method of displaying the contents of hard drives and the availability of programs. The functionality of Apple's GUI was emulated by Microsoft when it launched Windows.

In both the Apple and Microsoft paradigms, the look and feel of the GUI, as well as the resultant functionality of the machine, create the personality of the computer's brand. To a great extent, people, especially Apple customers, buy and remain loyal to a brand based on the GUI.

Mac OS   Generally speaking, art designers and illustrators, people who traditionally came from the print trades, favor the Mac. Because Apple once had an aggressive program to offer their computers at low costs to schools, many young people now entering the profession had their first computer experience on an Apple. The Mac OS is an easy OS to use, and because all the devices in the Mac OS have either been manufactured or licensed and rigorously controlled by Apple, the reliability of systems configured under this brand are highly reliable and rarely suffer the incompatibilities that lead to frozen computers and software crashes in Windows computers.

---

[2]Those of you who remember the recent introduction of the operating program, Linux, may have experienced this primitive stage before Linux achieved greater graphical sophistication.

Setting up, configuring, and reconfiguring a Mac is simple, requiring almost no knowledge of the computer's inner workings. Therefore, the Mac OS is a superb choice for those individuals that want to work with a computer, but who do not want to be burdened by the extracurricular tasks of learning computer technology. For most users, even those entering the complex world of computer animation, the Mac is the easiest system to learn and operate, but one must pay a price.

The drawbacks of the Mac are its cost and its speed. Apple computers are still, chip for chip, the most expensive computers to own. Many will also admit that the Mac is not the fastest computer when, to risk a pun, apples are compared to apples and a Mac is run in a side-by-side test with a similarly equipped Windows machine. The reason for this is that the high level of user friendliness of the Mac OS takes its toll in programming speed. However, new advancements in the Mac OS have addressed some of this and, of course, new levels of user friendliness may make the loss of speed a negligible issue for certain users.

One thing is certainly undeniable about Apple computers. The Apple customer is a very, very loyal customer. (See "Brand Loyalty to the Macs"). Apple must be doing something right to earn such trust from its 1 percent share of the global marketplace.

Windows   Microsoft Windows is by far the most popular OS, but I don't think one could say that brand loyalty has much to do with Windows' success. Windows is the OS customers love to hate. Shortly after each version is launched, one or more service packs are released, to be installed by the customer in order to fix or improve aspects of the version that should have been closer to perfection when launched. Often these improvements correct problems that cause crashes.

Until recently, the term "Microsoft Technical Support" was an oxymoron, but significant improvements have been initiated that I found impressive. Newer versions of Windows also include an automated feature that reports bugs back to the manufacturer.

To Microsoft's credit, it must be said that the Windows enterprise is vast and complex. Although Mac users are mostly to be found in the home consumer and graphic arts markets, Windows is applied by all business categories in practically all countries of the world.[3] At any given moment in

---

[3]About eight years ago, my company was contracted to translate just the logo and security tab for one version of Windows. Twenty-four target languages were used and the results filled five CD-ROMs.

time, multiple versions of Windows are designed for various types of users, and Microsoft plans even more.

The global market share of Windows was recently reported to be 97.46 percent,[4] leaving second place Mac at 1.43 percent and Linux in third place with .26 percent. Clearly, Windows is a mass product. It is designed for serious business computing for a global marketplace with a vast range of applications. As such, it is the perfect OS for 97.46 percent of the planet's computer users, and that's about what it is. You can do superb 3-D animation in Windows as well as use it to balance your checkbook, check the steam pressure in a nuclear power plant, or allocate U.S. grain deliveries to 70 million Russian farmers.

Because Windows is not brand specific in terms of hardware or third-party programs, any entrepreneur can design for Windows with a reasonable expectation of success. If you were a program developer for Alias | Wavefront, discreet, or NewTek, would you want to bet the farm on how many of the 1.43 percent of the available computer users will buy your product? That's not a test question. The answer is obvious. That's why it's getting harder and more expensive to buy third-party Mac OS software and easier and less expensive to buy Windows software. The same goes for new hardware.

You can hate Windows and throw darts at photos of Bill Gates, but it's hard to stick your argument to the wall when the wall is papered with the sales receipts of 97 percent of the world's computers. You might, however, think of walking down the hall where the remaining 3 percent of the world's computers, currently divided up between Mac OS and an admittedly growing number of Linux users, are warming up leftover lunch in the microwave. They'll listen, but Bill Gates is eating their lunch.

Linux It is important to note the latest starship to cruise into the fray of the OS wars: Linux. Unlike Mac and Windows, a corporation did not invent Linux for profit. Programmer, Linus Torvalds and a group of volunteer hackers from the Internet wrote (and still are writing) Linux from scratch.[5] Although Linux was experimental for many years, you could, and still can, obtain free copies of Linux and use them on any computer, Mac or PC, to achieve highly functional results.

---

[4]Amsterdam, OneStat.com, September 10, 2002, www.onestat.com/html/aboutus_pressbox10. html.

[5]You can learn more about the fascinating history of Linux and its current state of development, including useful links and Linus Torvalds's email address, at www.cs.helsinki.fi/linux.

The benefits of Linux are that it's free and employs an open architecture. If, for instance, you don't like the way Windows displays the contents of a drive, even if you're a sophisticated programmer, you can't change the Windows program. On Linux, you can change anything. You can even invent a new GUI (or any other feature, such as a handy **plug-in**) and offer it to the public for free or for profit.

Eventually, several companies did just that, creating first peripheral programs (called **drivers**) to run printers, modems, and other devices. Later, they developed highly functional GUIs that made Linux almost as easy to use as Windows or Macs and, in some cases, made it more reliable and less prone to crashes.

You can download Linux free or buy inexpensive CD sets that offer software and other advantages. Free downloads are available at **Slackware** (www.slackware.com), and low-cost CD sets can be obtained from distributors. Among these companies, the leaders are **Red Hat** (www.redhat.com), **Caldera** (www.caldera.com), and **SuSE** (www.suse.com).

IBM has recently jumped on the Linux bandwagon, a move that promises to produce significant results in the near term. With the financial muscle that IBM has, Linux may soon become a viable OS parallel with Macs for user friendliness and with Windows for business applications and mass appeal. IBM already claims to have scored major successes using Linux with such front-line companies as Mobil and Merrill Lynch.

It will be interesting to see if Linux rises to such prominence and, if in doing so, it remains the free, open platform that sophisticated computer users have asked for since the dawn of the microchip.

Brand Loyalty to the Macs

Not too long ago, I was hired by a certain computer company to create a **focus group** that would probe the loyalty of users to their brands of computers. A focus group is an assembly of 10 to 12 people who are paid $100 to $200 to sit around for 2 hours and talk to an interviewer about their opinions. Focus groups are the core methodology of **opinion research**.

This particular focus group was set up with three types of computers with identical specifications. The only difference between the computers was the OS and the CPU. The first computer was the fastest

Mac on the market, and the second was a brand-name PC running Windows with the fastest Pentium. The third was my client's computer, also running Windows, but equipped with a radically new type of CPU.

The focus groups were gathered into two classes: PC users and Mac users. For each group, we sat them down in front of all three computers and showed them some sample exercises: a complex filter job in Adobe Photoshop, a difficult render execution in LightWave, and a massive batch program in Adobe Illustrator.

We would start each focus group by explaining that we were going to perform several common tasks that demonstrated the speed of each computer. We would start with the Mac, move to the PC, and then finally move to the client's computer. Of course, we could have had three people press the start key simultaneously, but by staggering the starts we were emphasizing a bias toward the Mac and PC.

In each case, my client's computer completed the task significantly before the other two computers' often by more than half the time. The reason, of course, was simple. The new computer's CPU was the fastest on the planet, running at twice the speed as the other machines, as measured in **millions of cycles per second** or *megahertz* (**MHz**).

The PC was half as fast as my client's machine, because it was running the same OS at half the speed. The Mac ran about 15 percent slower than the PC, because it was running an OS that was producing a higher level of user friendliness than the PC and the extra routines involved in that process took a bit more time.

After the demonstration, we asked the group, "How much more would you expect to pay for this new computer than what you'd pay for either of the other two machines?"

The PC users unanimously answered with prices that were higher than the current average price for a PC. Many offered to buy the demo machine right out of the focus group. These people were seasoned graphics professionals, animators, video editors, and interactive media producers. They knew that time was money and that whatever the cost of the computer, they could amortize that cost on the basis of the time the computer would save.

These people had no brand loyalty. The fastest computer was the computer they wanted to own. Even after the session was over, the PC

(continued)

Brand Loyalty to the Macs *(continued)*

users hovered around the new computer, trying out various tasks and challenges from each of the installed graphics programs, as they muttered exclamations of awe and admiration.

The Mac users, however, behaved significantly different. After the demonstration, when asked what they would pay for the new computer, hardly a response was made. When probed further, the Mac users admitted they were amazed at the speed of the new computer. Some of the Mac users even admitted sadness that their Macs—especially the Mac we used that was, in most cases, a faster Mac than they could afford—could not perform as fast as the PC and certainly not as fast as our new computer. Like the PC users, the Mac users clearly understood that their Macs were costing them money in terms of performance speed (not to mention purchase price).

My clients wanted to test just how loyal the Mac users were. They offered to sell the Mac users the new computer for the same price as the PC. No, let me refine that statement: Even when my clients offered to trade our new computer for the Mac users' old Macs, every one of them refused the deal! "I just like my Mac" was the universal reply.

This is what manufacturers call brand loyalty. In its maximum manifestation, as demonstrated here, a customer will pay more for his or her favorite brand, lose time and money as a result of his or her purchase, and even admit to making an unwise choice, rather than change brands.

Choosing the Hardware

Once you've made the choice on an OS, your choice of computers follows suit. If you've decided to go with Windows or Linux, you have an unlimited array of hardware choices. You can start from scratch and build your own computer by purchasing a chassis, **motherboard**, **power supplies** (why not two?), memory chips (as many as you can afford—they're cheap), CPUs (again, two is better than one), a video card, hard drives, and a monitor.

Building Your Own Computer

Your key purchase is the motherboard, which can offer a wide array of features that once were available only by installing **peripheral cards**. Now motherboards have built-in **networking ports** (for connecting your computer to other computers), **modems** (for connecting your computer to a phone line), and **hard drive controllers**.

Visit your local computer supply shop. Some, such as **CompUSA**, **CDW**, and **Best Buy,** are national chains. Others operate as small corner stores (my favorite being **Computer Expert** on 46th Street and Lexington Avenue in NYC). Shop around and you'll find some interesting, inexpensive offerings.

The advantages of building your own computer are as follows:

- You get to learn the technology from the frame up.
- You can configure the computer to your specific needs.
- You save money (see Figure 3-3).

Figure 3-3
The author, delicately soldering a connection that will add an automatic backup power supply within an animation workstation. The proprietor with hands-on hardware experience will be able to save money and rescue an equipment failure faster than one who relies on outsourced technical assistance. "Learn the gut stuff."

The disadvantages of building your own computer are

- It takes time, especially the first time.

- You don't have anyone to call for technical support or warrantee service, although you can call the maker of each component.

Off the Shelf   Until recently, I always thought it was better to build my own computers, and I've built several over the years. Recently, however, the prices of fully configured computers have plummeted and I've grown weary of wiring.

Most computer manufacturers will offer you an assortment of preconfigured machines with a variety of features or the ability to customize the machine to your own specifications at a slightly higher price. **Dell Computers** pioneered the customization approach, and now others have followed Dell's lead.

*Although most animators want to buy the latest computer on the market, money is sometimes a limiting factor. If you're short for cash but want raw power, you should call each manufacturer's headquarters and ask for the department that sells **reconditioned computers** or **B-stock**.*

Computer companies, as warrantors of their equipment, must receive back into stock any computer that does not meet the guarantees of the product. When returned, that inventory is dead unless the computer company can fix it like new and resell it. This is called **reconditioning**. However, federal law prohibits a reconditioned computer to be sold as new, even if the computer just needed a new $15 keyboard. Once a computer is returned, it can never be sold again for full price. Thus, computer companies always have a way of selling their reconditioned computers. You essentially get a new computer, with a full warrantee (so you, too, can return it if it doesn't work perfectly), but you pay about 30 percent less.

B-Stock computers have been used by the manufacturer for demonstration purposes, such as trade shows. They may have some hours on them but are otherwise in fine working order and also come with a new computer warrantee. However, B-Stock cannot be sold for new. Therefore, the manufacturer sells them as B-Stock, which means you are free to negotiate a price based on the hours of wear the hardware has endured and how badly the manufacturer wants to unload the gear. Often a manufacturer will allocate B-Stock to a favored distributor or retailer. Although the retailer is legally bound to tell you the gear is B-stock, and most do so, certain retailers, both domestic and abroad, are not bound by laws or morals.

Both reconditioned and B-Stock hardware are good purchases if backed up by the original manufacturer. The best way to find such gear is to call the manufacturer directly and ask for these two categories of products. I have obtained excellent products of this sort from such top-line manufacturers as Sony, Dell, Compaq (now HP), and **Gateway**. I'd publish the phone numbers here, but they change frequently.

Just call main operators and use the terms B-Stock and reconditioned; you'll eventually be directed to a busy office where products come and go so quickly they can't keep a list online. Chances are you will be offered one list of products on Monday, call back on Tuesday, and find a completely new list. Keep trying until you get what you want.

A Note to Apple Customers As you've no doubt learned from the previous couple of pages, if you want to use the Mac OS, you simply have to go out and buy an Apple brand computer. The choice is rather simple. You just go to an Apple dealer (they even have Apple stores in malls now) and buy the model with the fastest CPU, the most RAM, and the largest hard drive. That's all you need to know. Also, you'll be able to easily install Maya and LightWave on your Mac. Install whichever software you like and get working.

Later in this chapter I discuss the peripheral devices you might want to acquire, such as a scanner, printer, or video acquisition card. You can also buy these from Apple Computer or from a group of manufacturers approved by Apple for compatibility with the Apple line. My explanation of these devices and, to some extent, certain third-party software can be applied to any OS.

Because it would be agonizing to research, verify, and, worst of all, update the validity of every product's compatibility for every OS, I decided to clump all the peripherals and ancillary products related to animation into one group without consideration to OS. I know I'll be correct for 97 percent of you out there. The rest of you will have to do a little homework to find out if the products will work on your Mac. I'm sorry if this causes your trouser pocket to catch on a doorknob or two, but hey, that's life, kiddo. You bought the computer that's easy to use and doesn't crash that often, so you have some extra time, right?

Good Brands

I'm often asked to recommend specific brands of computers and associated hardware. I have a policy about this that you should know.

First, I don't make millions of dollars writing books. It's pretty much a hobby that pays for itself. So if I write about an expensive piece of hardware, you can bet that the manufacturer of that hardware put the gear in my hands at little or no cost. I do go out and buy some hardware now and then, but guess what? I usually buy the same brands that are given to me to review. After all, I know them best of all.

Next, you might wonder, could I be persuaded to review a clunker product if the manufacturer sent me lots of free products? Are you crazy? Do you know how much time it takes getting to know a product before you can use it profitably? A clunker software could wreck your computer's entire installation, costing you weeks of work or, worse, a format wipe of the hard drive and a reinstallation of all programs. A dud computer could cost you even more time, especially if you load it with your favorite software and start to do deadline client jobs on it (my favorite way of pressure-testing any hardware). They say time is money. Way not. Time is more valuable than money, because lost money you can get back. Nobody will get away with stealing my time, no matter how many free pieces of junk they send me, and you can bank on that.

Finally, too many good brands are available out there that everyone should know about. These are the brands that have enough working capital to get their products into journalists' hands and that have technical support people who answer calls and have the answers to a hurried writer's questions—you know, companies that have it together. They might not be perfect in all ways. In fact, every product, even great products, need some improving, and that's what a good reviewer can do.

Computer Workstations

In response to the question about what brands I'd recommend, here are some of those products that were used as the test bed for this book.

Compaq/HP Often the acquisition of one great company by another results in a dilution of the brand, if not the qualities that made both companies great. Not so with this recent merger, or should I say, series of mergers under HP CEO Carly Fiorina. The guts of the best HP animation computers go back several years to the innovative microprocessor engineering of the *Digital Equipment Corporation* (**DEC**). DEC went out on an extreme limb to develop a new kind of CPU (the **Restricted Instruc-**

tion Set Computer or **RISC**) that allowed quantum leaps in processor speed.[6]

DEC, however, could not support its aggressive research and development with commensurate sales and became a takeover target. Compaq, one of the successful mass marketers of computers from the Dell and Gateway stratum of the industry, acquired DEC in 1999. The combination improved both companies, giving DEC's reduced but still-alive engineering team enough time to develop incredible rendering platforms. These new developments were used by such companies as **Blue Sky Studios** and **Pacific Data Images** to produce major animated films such as *Bunny* (the 1998 Blue Sky Studios' Oscar-winning short film) and *Antz* **(1998, PDI/ DreamWorks)**. *Star Trek: Insurrection* (1999, Paramount) also featured animated effects by Blue Sky Studios. At the same time, the popularity of Compaq and its legendary **Alpha RISK** kept the cash flowing, at least for a time.

Eventually, even Compaq suffered proportionately with the other computer companies in the bloodletting that crippled the dot-com revolution. Compaq was acquired by an equally beleaguered HP in 2001, and currently the merger is being called "the New HP," although certain HP facilities and products still use the Compaq name.

As of this writing, the combined forces of DEC engineering, Compaq marketing, and HP reliability are producing some of the fastest and most reliable computers for animators. Their key vendors have continued to release animated Hollywood blockbusters like *Ice Age* and *Shrek*.

My current favorite configuration is a **Compaq EVO Workstation W8000** with two Intel Xeon 2.80 GHz/512 KHz processors, at least 1GB of RAM (configured as two 512KMB SIMS), and an **nVIDIA QuadroFX 2000 AGP** (8x) video card with 128MB of RAM (see Figure 3-4). Because the W8000's motherboard comes with an integrated, dual-channel Ultra 160 SCSI controller, I prefer booting from an 18.2GB Ultra 160 (15 Krpm) SCSI drive. For convenient CD accessibility and backup, I install a 4.7GB *DVD+ Recordable/ReWritable* (DVD+R/RW) drive in addition to the standard 48X

---

[6]Quick example: As I write this in 2003, the speed of CPUs has doubled every year to a maximum of 2.2 GHz as the current best. However, a 700 MHz Pentium CPU is still considered to be pretty fast and is the most common CPU speed sold today. DEC introduced the first 750 MHz CPU at the National Association of Broadcasters convention in May of 1998. True, it had a mini-refrigerator screwed to the chip, but had DEC continued apace with the current industry, well, you do the math. Would they now have a 12 GHz CPU?

Figure 3-4
The nVIDIA QuadroFX 2000 video card, which was installed in the author's Compaq EVO Workstation W8000. Notice the hefty, dedicated cooling fan?

CD drive. I round out the iron with a pair of 17-inch **Planar** (www.planar.com) flat-panel monitors and a pair of Seagate 180GB Ultra 160 SCSI drives for rendered frame storage.

This configuration will tip the solvency scale at just about $10,000, but this is a machine that leaves nothing to be desired as an animation workstation. For more information, go to www.hp.com/go/evoworkstations, a web page that lets you build your own system and see the price rise and fall.

IBM I have no reservations whatsoever about recommending IBM for mission-critical work. With the demise some time ago of **Intergraph Computers**, which were overengineered (in a good way) and tech supported beyond imagination, IBM attains my first choice as the computer company you want on your side when a project busts open like a rusty tank of gasoline and the client is playing with matches.

My IBM **IntelliStation** for some unknown reason recently failed to be able to access the Internet from its onboard Ethernet port. A minor annoyance, but it was Christmas Eve—I kid you not. I called IBM tech support and got immediate, live technical support without a lot of "touch tone your serial number now" idiocy. These guys knew I was asking for help and they were there and on it. The problem could not be resolved over the phone and I was going on vacation the next morning. Not under pressure, I told the tech guy I would get back to them after New Year's by accessing my case number.

While I was away, my assistant got me on my cell phone at the beach. "The IBM guy dropped by and fixed the IntelliStation," he said.

"What?" I replied, "On December 26th?"

"Yep. He said he was just passing by and had the parts."

It's nice to know that companies are still out there that can afford and are equipped to deliver this kind of reliability. I'm just glad I can tell you this; certainly, a lot of little animation companies like ours live or die based on guys like that IBM technician.

Do I also need to tell you that IBM is the oldest computer company in the world? I remember my aunt coming home in 1957 all gleeful because she got a job as a keypuncher at IBM. The computer term **bug** comes from IBM (attributed to a cockroach that crawled across a high-voltage switch deep in the guts of a computer that took up several rooms and couldn't compute as fast as a Casio wristwatch). HAL, the name of the renegade computer in the Kubrick film, *2001: A Space Odyssey*, came from backing up each letter of IBM. (Legend has it that IBM would not allow Kubrick to use their name for free, although PanAm did, but where's PanAm now?)

IBM has certainly had its ups and downs. Just as it was the first computer company, it was, for ages, the only computer company. As most monopolies, it grew fat and narrow minded. Steve Jobs and Bill Gates completely blindsided IBM. The first IBM PC, when it finally did arrive, long after Apple had been around, wasn't even developed in house, but by a band of freelancers holed up in a Florida hotel.

Throughout the '90s, however, under the brilliant helm of Lou Gerstner, IBM got turned around. It found out where the beef was located and slowly started carving out its share. Unlike many other computer companies, today IBM is wisely dicing up the market in the vertical direction by creating entire product lines and service channels dedicated to specific industries, such as broadcast news. They are also involved in the horizontal market, by offering total solutions that include machines, software, personnel, physical plants, and service bureaus.

In the IBM IntelliStation Z Pro Mini Tower configuration I used during the writing of this book, I specified two Intel Xeon 2.80 GHz/512K processors and 1GB of RAM. The video card was an nVIDIA Quadro FX 2000 with 128MB of RAM, the latest and best offering from this manufacturer. I am biased toward nVIDIA because they are currently making fast cards that are compliant both to the OpenGL and Direct X **application programming interfaces** (**APIs**). Although Direct X has been a utility employed by gamers to render fast 3-D emulation for many years, it has recently begun to pass OpenGL as the accepted API and I believe will soon be the superior API for animation. Currently, LightWave and 3ds max support immediate toggling between the two APIs.

IBM makes superb hard drives at a fair price, so I went with their default boot drive and upgraded the configuration with two 73.4GB 10Krpm Ultra 160 SCSI drives for frame storage.

After mounting the dual flat-panel monitors for the Compaq, I had no more room for monitors, so I chose just one IBM T860 18-inch flat-panel monitor for the IBM. Like HP, IBM offers an interactive hardware configuration page where you can choose your options and get an instant price. This computer slithers like an Alabama blacksnake into a tight little nook beside my left knee for just under $8,647.50 plus a video card.

Let's Consider Laptops

While I'm writing about IBM, I'd like to introduce the concept of computer animation on a laptop. Don't get me wrong here. A laptop will not outpace a deskstation in the foreseeable future. Although I hesitate to use the word never, you could argue that whatever can be put in a laptop will first be developed for a desktop, but who knows?

A laptop would not be a choice machine for rendering animation (although I will admit to having plugged them in to a render farm in a pinch), but it is a superb platform for really creative thinking. You can easily model characters, lay out scenes, and even execute low-polygon previews on a well-equipped laptop. For example, I have all three animation programs, LightWave, Maya, and 3ds max, open right now on my laptop as I write this.

Here's another little tip for you executive types out there who are reaching that age in your life (you know who you are) where you're considering the midlife change of career. Many of you work for corporations that try to compromise your home time by "giving" you a laptop so "you can be completely mobile." I think it's fair to use the computer at home a little for your personal life enhancement, don't you? So when the boss asks what kind of laptop you need to do your work "on the road," ask for a big mother that can handle animation. Then you can cram the PowerPoint work into your coffee break and learn animation on the plane to the Tucson sales meeting.

That's where IBM comes in again. I called their animation hardware specialists for recommendations and obtained an A31p Thinkpad in late 2002 to use especially for writing this book. I will admit it is a good bit heavier than my T21, on which I wrote *The Desktop Video Studio Bible,* but it's worth the extra work of heaving the backpack around. This model comes with an Intel Pentium 4, a 1.9GHz processor, a 36GB hard drive, and an ATI Mobility FireGL 7800 video card. I added 512MB RAM to the standard 256 and souped it up further with a Lithium-ion battery, a 10/100 Ethernet adapter, a Bluetooth wireless network adapter, a DVD-to-DVD/CD-RW insertable drive, and, the coolest device ever, a pop-out numeric keypad. Even the CPAs on my commuter train look up from watching the grass grow when I pop out the number pad and start punching in shader parameters. An optional DVD burner enables the user to output sample playbacks from the field. These optional items may be removed from the sides of the unit and replaced with more conventional devices, such as a floppy disk drive, or be left out completely to reduce weight (see Figure 3-5).

The price for this wicked little laptop is about $2,600—about what the boss should expect from a hard-working manager like you, who runs off to Tucson and leaves his little kids with the nanny at a moment's notice. Yeah, right on.

Cutting Corners

It wouldn't be responsible of me in this post 9/11 era of tight budgets to prance around with the best and latest brands of hardware, knowing some of you out there are scraping your last dollars together to get into animation. The computers I've recommended here are clearly the pinnacles of the trade, and although you should aspire to own one, you may only be able to squeak into the business with something you cobble together from an Enron auction or from your parents' largess. I feel your pain.

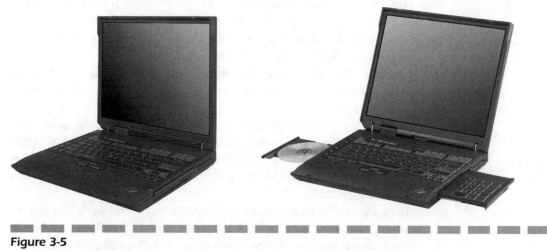

Figure 3-5
The IBM A31p with and without the pop-out numeric keypad and DVD/RW drive

Here are some cost-cutting techniques employed by entry-level anima-tors and just plain skinflints.

You could substitute the Intel chips with lower-cost AMD Athlon equiva-lents. They work fine. You could buy B-stock from the same reliable com-panies mentioned earlier. You could buy used gear or hardware based on CPUs that are a generation or two older than the current cutting edge, and, in the final analysis, you can use just about any computer around to do animation. The modeling might be a bit slower, and the rendering way slower, but when you're young, time is cheap. Also consider doing the modeling and staging on your home computer and then renting time on a larger network to do the rendering. Offsite rendering is often an option used by the best professionals. Who wants to keep all that hard-ware around between jobs anyway?

The Fastest Iron on Earth?

Once upon a time, 3-D animation was considered the industry that made the geatest demands on computers. You might be interested to know, that 3-D animation is no longer the industry that demands the fastest comput-ers. The industry that does, however could be a place you might want to work.

Steve Briggs, who sells large computers for HP (he came up the path from DEC and Compaq), has moved his focus recently from Hollywood to Detroit. "Car makers are getting into extremely high-end rendering," says Steve, "and it's not just to model new cars, but to examine how everything in a car works under various real-world simulations."

Steve adds, "Boeing is doing the same thing. Building virtual airliners and then seeing how every little bit works before they commit to making anything in a factory."

Elaborate simulations that challenge the limit of what can be known are the main focus of fast computers today. The experiments at the Lawrence Livermore laboratories using superclusters to determine the behavior of atomic weapons during atomic weapons test bans were part of how such simulations originated.

Perhaps we can learn a few things from the guys who work in the new simulation industry. In looking back to the days when he sold computers to animators, Steve refines the requirements of 3-D animation hardware when he adroitly states that current 32-bit processors, as opposed to the more expensive 64-bit models, are more than adequate for today's media producers. "Animators already have enough processing speed in the CPUs. It's **throughput speed—memory bandwidth**—that's crucial to animators now."

Memory bandwidth, expressed in bytes per second, determines how fast the components of a rendered frame and finally the finished frame can be transferred through the computer to its final resting place on a hard drive. A high rendering speed is obviously achieved when a separate CPU renders each frame in a sequence, and many CPUs are combined in a render farm. But, if each CPU cannot obtain the data to make that frame or return that data to storage as fast as frames are rendered, the bottleneck that results does not affect CPU speed but *bus* speed. Today it is bus speed that rules.

Steve's fastest computer, a Marvel EV-7 Quad, for instance, uses four Alpha processors running at 1.25 GHz, but the memory bandwidth is a staggering 25GB per second. Compare this to the leading EVO system bus running at 9MB per second and you have an idea where the priorities are heading in supercomputing. It's possible that tomorrow's leading animators may well be buying second-hand chassis from GM and Boeing.

Video Display Cards

If you are investing in a computer for 3-D animation, it pays to buy one from a *value-added reseller* (**VAR**) who has experience in the animation field. Such a reseller can recommend the latest and best hardware and software. One such recommendation is the choice of a video display card for your computer. The video card is a peripheral device installed inside your computer that supplies signals to the computer monitor. Video cards are nearly always supplied with any computer you buy, but the requirements of 3-D animation raise the level of capability that your video card must maintain.

Even generic computers are configured these days with rather hefty video cards, because video gaming is one of the tasks most home computers are expected to perform. Like computer animation, video gaming requires the capability to display colorful, shaded, 3-D forms quickly on a monitor, but most consumer video cards are not adequate for the kind of work you will be doing in animation.

In a workstation, the key to speed is being able to evaluate your work as soon as you enter commands. You might think, "Well, isn't that what all computers do?" No. All computers and video cards cannot take the 40,000 polygons of a galactic battle cruiser; position them; apply texture, lighting, and atmospherics; and throw them onto a monitor as fast as you can decide that the ship should be pointing toward the enemy instead of running away. So, on a computer that is not adequately equipped, if you punch in a 180-degree Y-axis adjustment, your screen will go blank. About 10 minutes later, it might start drawing the ship polygon by polygon. During the wait, your deadline keeps moving, and your ideas for the next change sit, unexecuted, maybe even forgotten. You get the point. You need instant playback of your creative input.

The way computers provide this instant playback is by *emulating* the look of your work. This word emulate is really important. When you are designing an animation, your software employs an algorithm that translates the complex display attributes of your scene into simplified emulations of the scene, which are composed of vastly reduced data. This reduced data is employed by the video card to show an emulation of the scene you are creating. Employing onboard memory and specialized circuit chips known colloquially as **geometry engines** your video card creates an emulation. The technology of this process is not standardized, and without reading extensive technical reports, it is difficult to match oranges to oranges in understanding the process.

Some of the factors you want to examine are onboard RAM capacity and **floating point rendering** capabilities. Floating point precision has recently been upgraded from 32 bits to 128. This has raised the capability of the board to emulate colors, from 256 variations per color in 32-bit floating point to several million in 128 bit.

A good test, if you can arrange it at a computer store, is to load the same complex scene on two identical computers fitted with the two different video cards you want to test. Then time how fast the complex scene is emulated on each computer. Do you think you can do this? Doubtful. Even as a tech writer, I've only had the opportunity to do it on a handful of occasions, and as soon as I had publishable results, the boards were upgraded and the marketing departments issued a new load of mumbo jumbo. I rely a good deal on Internet user comments, supported by in-house results.

Generally, the faster the video card and the greater the onboard memory, the faster and more accurate the emulations will be. That's why it pays to have a fast, high-memory video card; it provides faster and more realistic emulations of your work so you don't have to wait or guess what the end result will look like. However, keep in mind that you are only looking at an emulation. You have to actually *render* the frame to see the final result. Because the render is created pixel by pixel in your computer, not in your video card, it takes longer to view, but once viewed, it is the precise result of your work, not an emulation.

Video card development is one of the fastest developing areas of our business. By the time I finish this sentence, a new card specification will be issued. Be sure, when investing in a high-end video card, that the card is compatible with your computer and with the 3-D animation software you intend to use. Here's what's hot now:

- **3Dlabs' Wildcat4**   Designed to work with an AGP8x graphics bus, this card offers a choice of 384 or 256MB of onboard RAM, dual-monitor support, floating point precision, and synchronization capabilities with several workstations (this is called **genlock**).
- **ATI's The FireGL X1**   Designed to work with both AGP4x and 8x, this card supports 256MB of onboard RAM, dual-monitor support, 128-bit floating point precision, and Linux compatibility.

Scanners and Printers

Animators will find it necessary to obtain both a high-quality scanner and a color printer. The scanner is a device that enables the animator to convert **flat art** samples, such as sketches or photographs, into digital form. The printer enables the artist to output flat art designs for use in storyboards, client presentations, and other forms of noncomputer distribution. The printer is also essential in generating business opportunities for the animator, affording her a low-cost means of creating promotional literature that may be mailed to prospective clients.

Scanners are useful in a variety of ways. Most animators begin their work, especially characters, by sketching with a pencil and paper. Often, animators create scenes from paintings and photographs they see or photographs they take. My own shop has even scanned 35mm motion picture film clips and photographic sequences to capture the realistic motion of animals and humans. Each of these real-world, flat art samples can be scanned into your computer and used as background references in animation software programs. Sometimes the flat art will be used as a reference guide, whereas other times the art may be incorporated directly, such as a background. Whatever the use, scanners are an indispensable tool for the animator.

Both scanners and printers have declined significantly in cost since the early days of computer animation, allowing a scanner to be economically added to each artist's workstation. Simply as a means of saving desk space, one or two high-quality printers can be placed in a central location and networked to the workstations. It is a simple matter to send art to the printer and collect it at a convenient time. A scanner, however, requires constant manipulation during acquisition (see Figure 3-6).

Both scanners and printers are evaluated according to resolution, as measured in **dots per inch** (**DPI**), and speed, as measured in how quickly an average color or black and white page can be produced. A typical color printer, networked to several computer animation workstations, can assist in producing storyboards, client evaluation sheets, and promotional materials for the animator (see Figures 3-7 and 3-8).

Audio Capabilities

A studio that employs computers with only the most minimal audio playback capabilities can accomplish a significant amount of animation. How-

Figure 3-6
A typical scanner for digitizing flat art into computer animation. This model, the HP Scanjet 7400c, offers 2,400 DPI and high scanning speeds.
Price: $499.

Figure 3-7
If you can live with output on paper no larger than 8.5 by 12, the HP Deskjet 6127 is an excellent choice. It provides 4800 × 1200 dpi in color and prints at 11 color pages per minute.
Price: $249.

Figure 3-8
If you want to print on larger paper, you'll need the HP cp 1700ps, which handles paper up to 13 by 19 at up to 14.5 pages per minute, with a resolution of 2400 × 1200 dpi. Price: $499.

ever, the ability to import, create, and export high-quality audio will enhance your workstation's profitability.

In about 30 percent of the animation work I have done, especially for corporate clients, the creation of audio tracks for the contracted animation was a competitive advantage. In most cases, this led the client to increase the budget specifically to include audio tracks, theme songs, and effects I offered as part of the deal.

Designing quality audio for animation requires three essential elements: a computer sound card, sound-editing software, and sound creation (or synthesizing) software. A recording booth, an audio mixer board, and stock audio playback devices are also useful to have installed.

The Sound Card   Although marginal sound quality from an off-the-shelf computer is still acceptable in some areas of digital media creation, the standard installed sound card is not recommended for animation work, where your creations will sell for thousands of dollars. If you're going to offer professional audio with your 3-D animations, you should invest in a professional-level sound card.

Currently, some easily distinguished features are found only on professional cards, such as **XLR** (balanced) line connectors, multitrack capabilities, and a high rate of data transfer between the card and the computer (called the **bus rate**, which is measured in bits). The trick is to get a card that is the same bus rate as your computer. Assuming you will be using a computer that is already equipped with the highest bus rate, your card should be compatibly fast. If you will be executing compositions with several tracks of sound, your card will need to take time to render its results if it is not equipped at the hardware level to record and play the various tracks in real time.

Due to rapidly advancing technology, it would be unwise for me to quote the number of tracks and bit rates you should look for. Check what's available for the average workstation and then search for cards with significantly higher specifications. Certainly, you want more than two tracks and a 32-bit bus. The better you buy, the more you will be able to do, now and in the near future.

Many professional sound cards also come bundled with sophisticated sound-editing software, albeit usually the "lite" versions. You can also distinguish a professional sound card by its **sampling frequency rate**, which will be considerably higher than the current consumer models. The frequency range used to be a factor, but now nearly all cards have ranges that far exceed the human ability to hear (20 to 20,000 Hz). As I write this, the current demand in sound cards is the capability to render the various versions of Surround Sound, especially the current Holy Grail of Dolby 5.1 (which employs six speakers to emulate theatrical motion picture quality).

Many professional sound cards come with elaborate breakout cables that attach to the sound card at the rear of the computer. Such cables offer multiple channels of XLR input and output connectors, as well as an assortment of digital audio connectors such as the *Advanced Encryption Standard/Enterprise Backup Utility* (AES/EBU) and the *Sampling Probability Density Function* (SPDF). These connectors are useful if you have any gear such as *digital audio tape* (DAT) players or CD players that offer similar digital connectors.

An interesting example of a professional sound card is the **Nuendo 96/52** from **Steinberg**. (As of this writing, Steinberg [www.steinberg.net] has recently been acquired by Pinnacle Systems [www.pinnaclesys.com].) The Nuendo 96/52 takes a different approach from most sound cards by having the drivers built in to the card (instead of relying on drivers being installed into the CPU). This results in all data transfers being shifted to and from the PCI bus directly, without involving the CPU at all. Supported sample frequencies are 32, 44.1, 48, 88.2, and 96 kHz. The card presents zero CPU load, even when using all 52 channels (26 I/O simultaneously). Up to three cards can be installed in one system. Maintaining synchronization between them is as simple as connecting them all to a common device. The Nuendo 96/52 retails for about $500. When choosing a sound card, ascertain if your growth plans will eventually include a full nonlinear video-editing platform, within which sound capture peripheral services may be included.

Sound-Editing Software

Excellent, economical sound-editing packages are available for both the Mac and Windows. For the Mac, the **Digidesign** (www.digidesign.com or 800-333-2137) **Pro Tools** assortment of products is the most popular software for recording, mixing, and editing audio. Digidesign has also launched a line of Pro Tools products for Windows-based hardware as well, extending the brand's customer base. Pro Tools products are available in many variations, some as turnkey boxes, others as software only.

A wide range of plug-in utilities (both by Digidesign and third-party vendors) is also available. Low-cost systems, such as the **M-Box**, are available for as little as $500, whereas the workhorse of the product line is **Pro Tools|24 MIX** (beginning at about $8,000 for a turnkey system). MIX offers dedicated audio recording and mixing capabilities that provide all the professional necessities of a dedicated sound facility (see Figure 3-9).

Another sound-editing solution designed for Windows is **Wavelab**, produced by Steinberg/Pinnacle Systems. Wavelab is a software-based product that offers an inexpensive, easy-to-use interface similar to Pro Tools's time-line design. Combining easy-to-use stereo editors with a wealth of mastering

Figure 3-9
Ace mixer, Jeremy Goldsmith, at work in his Pro Tools-equipped sound studio

features, WaveLab offers powerful editing tools and an array of high-quality virtual-effect processors. For those interested in using their computer to burn CDs for personal use, comprehensive CD burning capabilities, real-time audio file analysis, and batch processors further add to its functional depth. Steinberg's latest version also supports samplers, **audio stream input output (ASIO), wavelength division multiplexing (WDM)**, and a vast range of file formats.

Novices can use most of WaveLab's functions straight out of the box without much documentation reading. The software is offered in two versions: Wavelab Essential ($299), which is a bare-bones form of the other version, Wavelab ($499). Both are quite adequate for the average animator, and an upgrade kit is also offered for the difference in price between the two versions (see Figure 3-10).

Amplifier and Speakers

Consider how important sound is whenever you watch animation. If you are planning on having your clients visit your facility to see your work,

Figure 3-10
A montage of some of the Wavelab screens that assist in analyzing and editing sound files for their eventual export to animation in videos, CDs, DVDs, and games

get yourself something better than those cheesy plastic speakers that come with most computers. You want a professional amplifier and studio speakers.

Fortunately, you don't need anything really powerful, because the sound only needs to be loud enough for you and your clients to hear while seated at the workstation. You're not equipping a club lounge or Yankee Stadium here. Choose an inexpensive amplifier with about 50 watts per channel. The speakers, however, should be the best you can afford, such as those found in high-tech audio stores or professional multimedia supply houses. The general cost range of the amplifier and speakers should be $150 to $300 for the amplifier and from $100 to $500 each for the speakers.

Electronic Pen Devices

A popular though not strictly necessary tool in animation is a device called an **electronic pen**. Electronic pens enable you to substitute your mouse for a pen device. You use the pen just like any pen, pencil, or brush, depending on the software and how you set the defaults. Obviously, if you come from a traditional art background (or if maybe you think the mouse was never designed to, well, draw), a pen device is a lot easier to use than a mouse.

Electronic pens come in two basic types. The ones that require a dedicated surface or tablet on which to write are called **pen tablets**, including the pen itself. **Screen pens** are designed to write directly on your monitor screen.

One company, **Wacom** (pronounced "WAH-come," www.wacom.com) is the leading supplier of both types of pen input devices. The most common product from Wacom is the **Intuos** pen tablet, available in a wide range of sizes from 4 by 5 inches ($200) to 12 by 18 inches ($750). The Intuos line features a wireless pen that serves as a cursor and input device for whatever software you are running when it is brought to the surface of the tablet. Several color-coded pens configured for different artists can be used on the same system if more than one operator needs the tablet.

Wacom has also recently released a new screen-pen system, the **Cintiq**. This system is quite exciting, combining flat-panel convenience with the hands-on touch and feel of an easel, and it includes a pen that can write directly on the flat panel's surface. Two models are available. The Cintiq 15x ($1,899) features a 15-inch screen, with 9 by 12 inches of drawing area. The Cintiq 18sx ($3,499) features an 18-inch screen with an 11.3- by 14.1 active area. Although quite expensive, the Cintiq line stands as a paradigm of the best way to draw on a computer (see Figure 3-11).

For those artists wanting to write directly onscreen but not having thousands to spend for a Wacom Cintiq, another option is **FastPoint** (www.fastpoint.com), a wired pen device that connects to the computer. This device works off the light from the monitor and is a bit unreliable in exceptionally dark areas of a design, but otherwise it works quite well. I use FastPoint on my everyday computer just to avoid carpel tunnel syndrome. Prices for various types with different button configurations run about $400, including the computer interface.

Optional Tools

In my first book, *The Desktop Video Studio Bible*, my premise was that because you were creating your primary medium (in your case, animation) on a computer, and because all other media (writing, web sites, DVDs, CDs, and video) could also be created on the same computer, why not create projects in those media too? Animators are a concentrated, specialized group of artists. It is not easy, and perhaps even harmful, for animators to spin out on tangents away from their core expertise. And yet, here you are sitting on a machine that can make you more money. Maybe you can hire someone to do the other media when you go to bed? If this premise is worth considering, here are some optional tools (and profit streams) you might consider.

Audio Creation

It should already be apparent that adding sound to an animation is a vast enhancement. Watch any of the spaceship battles in the *Star Wars* saga with the sound turned off and you'll understand. Find a telephone pole with cables holding it steady to the ground. Hit the cable with a hard object like a hammer and listen. That's the laser blast sound, which was essentially created the same way. Record the sound, add it to the right scene, and dull becomes exciting. Sound creation for animation is as much fun as making the animation itself, maybe more because the sound adds exponentially to the pleasure of seeing the animation. Creating your own original audio tracks to accompany your animation will also add a competitive edge to your company's offerings.

Film and television producers, as well as large advertising agencies, however, will commission animation projects as ***mit out sound*** (**MOS**) projects. You need never supply the audio elements, although you may often have to synchronize your animation to someone else's tracks. That's why you may need a good sound card in your workstation.

Corporate clients and small advertising agencies usually do not look for a separate source when it comes to an animation's sound effects and music. Often they will be delighted to hear that you can supply both sound and picture. In such cases, you can either commission to create the sounds or do them yourself.

You would be surprised to learn how simple a sound can be and still do an effective job. In fact, most novice sound designers make the mistake of **overtracking** their compositions, that is, putting too many elements in. Start simple, test the result, and stop the second you are satisfied (well, chances are you won't stop, so at least save a version you can return to when you realize you've overtracked).

To make your own music and effects tracks, explore the capabilities of a program called Acid Pro, yet another innovative product from **Sonic Foundry**. Acid Pro is a loop-based music creation software package for Windows. This product comes with a few CDs full of sample clips. You can give these clips a timeline, so they can be easily synchronized to your visuals, and you can then loop and modify them. Sonic Foundry offers dozens of these CDs grouped according to the style of music you want to create. Because the samples included in Acid Pro are free of copyright restrictions and because the remaining creative work on the music will be done by you, the resultant tracks are considered yours to copyright and own. You can also capture your own sample sounds and insert them into the timeline too, making sound effect backgrounds, such as a jungle atmosphere, that perfectly match your visuals (see Figure 3-12).

For animators who really need a break from sanity, take a look at **Reason** from **Propellerhead Software** (www.propellerheads.se) (about $280). Reason, available for both the Mac and PC, is innovative software that replicates a complete sound-synthesis and mixing workstation on your computer's screen (see Figure 3-13). The interface replicates a 19-inch equipment rack of components. Start with an empty rack and click the menu of components that includes a synthesizer, a drum machine (called Redrum), a mixer, a reverb unit, an equalizer, and a timeline. You can flip the rack around to reveal the back where you'll find a patch bay. Grab the patch cords with your mouse and replug them anywhere you like (the colored patch cords even wobble as you move them about).

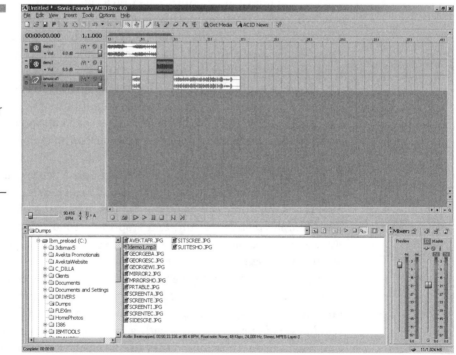

Figure 3-12
An Acid Pro session screen showing the timeline and a samples library. Acid Pro retails for about $200; additional CD libraries of loops sell for about $60 each.

Figure 3-13
Reason's design replicates the sound engineer's environment accurately with dials, switches, controls, and even taped names on the mixer pads. This software is all you need to build exciting tracks for your animations or games. It retails for about $280 for both Mac and Windows versions.

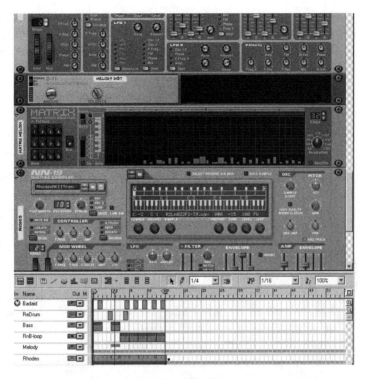

This software is best operated with a **Musical Instrument Digital Interface** (**MIDI**)-compatible, piano-style keyboard device. If you know how to play a piano, great, but you really don't need such experience to knock out simple melodies and enhance them with this software. You can also capture and process loops, beats, and samples from other sources using Reason's sister programs **ReBirth** and **ReCycle**. I know of one aspiring computer animator who fell in love with Reason and built a fairly good business burning loops for freestylers and poets. He hasn't turned out an animation in months.

Stock Music

One of the easiest ways to get music for your animations is to buy them from a **stock music company**. A stock music company acquires thousands of cuts of music with full rights for all media worldwide. You can then buy limited rights, such as nonbroadcast, educational use only, or Internet, as you need them. The fee you pay can be as little as $75. If you buy enough licenses, the stock music company will often lend you entire libraries of CDs to keep in your studio or entitle you to download music clips from a central Internet database. Some of the better-known stock music houses are Omni-iMusic, Killer Tracks, ProMusic, and FirstCom/MusicHouse.

The Recording Booth

A small **recording booth** is useful for recording the voices of animated characters, narrators, and, to a limited degree, simple music such as a guitar, harmonica, or drum. Although you may want to use a full-blown professional recording studio for commercial-level work, a simple, in-house recording facility can be useful for recording scratch tracks and for experimenting with various audio alternatives before you commit yourself to hours of laborious animation work.

The recording booth can be a small room or closet lined with sound-absorbing material such as corrugated foam on the walls and ceiling. The floor should be carpeted and the space should be equipped with a non-swiveling chair or stool and a music stand for a script.

Obviously, many thousands of dollars can be spent on a recording booth in terms of achieving perfect soundproofing, providing a window between the talent and the engineer, and offering adequate ventilation. Such studios rent for upwards of $200 per hour. For convenience, however, nothing beats your own recording booth. An adequate booth, with full equipment, should cost just a few thousand dollars to equip and repay your investment many times over.

SoundForms International (www.cinereps.com/soundforms.htm) offers inexpensive, collapsible recording booths of various sizes. They are composed of plastic tubing, fabric, and rollup foam panels that can be easily stored in the trunk of a standard size car. In a matter of minutes, you can assemble a Single Sound Booth, measuring $3\frac{1}{2}$-feet wide by $3\frac{1}{2}$-feet deep by $6\frac{1}{2}$-feet high, either in your workspace or on location. Other sizes are available as well (see Figure 3-14).

In addition to the recording booth, you'll need a few other items, mainly a microphone, headphones, and the wires to connect them to your computer. A professional microphone will always have one feature to distinguish it from a consumer model: an **XLR-type connector**. An XLR connector is a metal or plastic cylinder, measuring 5/8 inches in diameter, with three connecting pins. This is also called a **balanced line** connector.

The microphone is placed on a mic stand, which is often equipped with a circular cage-like device, which holds the microphone suspended from elas-

Figure 3-14
The SoundForms
International
Single Sound
Booth is ideal for
field or office use.
The model is
shown with
optional
microphone
mount for $1,200.

tic bands or foam rubber to isolate the microphone from extraneous vibrations. A balanced XLR type of microphone cable connects your microphone to your mixer.

Your recording booth must also have a method by which the voice talent can hear the sounds you are recording. This is supplied by a pair of headphones (one set for each performer) that is connected to the **monitor output jack** on your mixer or through a small **amplifier** connected to one of your computer's audio outputs. During your recording session, you should send a signal from the mixer or computer back to the voice talent, so the talent can hear the recording at a comfortable level.

The Audio Mixer

An **audio mixer** is usually required if you are going to feed more than one form of audio into your computer or if you are recording voices. The mixer enables you to connect each audio device to its own channel and you can determine how loud that device will be recorded in relation to other devices.

Audio mixers are usually priced and described in terms of the amount of inputs and outputs they feature. You might only need one of the smallest

Figure 3-15
A typical economical mixer for professional use. This mixer features 12 inputs and 2 outputs. Each channel features dials to adjust panning, equalization (EQ), filtering, gain, and effects. This example, the DFX-12 from **Mackie Designs** (www.mackie.com), goes for approximately $400.

mixers, which are usually four by two (also designated as 4:2). This means the mixer has four inputs and a pair of stereo (left-right channels) outputs (see Figure 3-15). Other popular configurations are 8:2 and 12:2. Anything more elaborate is probably overkill. Of course, the mixer should feature a **monitor jack** where the voice talent's headphones can be plugged in.

Some mixers also feature an inexpensive microphone or a place to connect such a microphone that enables the mix engineer to speak back to the person in the recording booth.

This eliminates shouted conversations such as, "Okay, Bill, give it a rest." "What? Why am I a pest?"

Microphone

For voice-over work and many musical applications, you should select a **condenser** microphone. A certain type of condenser microphone, called a **cardioid**, provides an added benefit for use in small recording booth work (see Figure 3-16). This kind of microphone has a heart-shaped zone of sensitivity. The top and sides of a cardioid microphone, whereyour voice would be projected to, are most sensitive. At the bottom, where your script would

Figure 3-16
A typical cardioid condenser microphone and its sensitivity pattern. Notice how the sensitive area is heart-shaped, providing a dead zone for selective recording in tight quarters. This model is the C2000B from AKG.

be rustling, the mike is less sensitive. In music recording, this pattern can also be employed usefully by pointing the sensitive zone toward the instrument to be recorded, whereas the dead zone can be facing another instrument or voice you might want to eliminate and record separately.

Headphones

The headphones in your recording booth are necessary for allowing you to hear the recording being made (see Figure 3-17). If the recordings are voice-overs, actors must hear how they sound in order to produce the best vocal quality. In music recording, the musicians can hear the session with various levels of backup instrumentation mixed to the musician's preference. Musicians need to hear the studio mix in order to tune their performances to it.

Looping Tools

In creating character animation, you have the choice of creating the animation and having the actors match their voices to the characters' lips, or vice versa.

The former choice is called **looping** or **Automated Dialog Recording (ADR)**. The term comes from the old days when audio was recorded on magnetic tape and film. A loop of tape was made that matched the length of the

Figure 3-17
The recording booth headphone should have ample cushioning to isolate the talent's ears from external sounds. The talent can then concentrate on the mix fed through the headphones. This model, an HD570 from Sennheiser, is listed at $130.

segment to be recorded, and the actor simply kept saying the words while watching the film until a perfect synchronization was established. The engineer could, of course, punch the record button off and on to capture one word or phrase until the entire sequence was perfect.

An advantage of looping is that your actors get to see the characters they are playing and adjust their performances accordingly. If you are planning on looping your animations, you will need two essential tools. First, a monitor must be installed in the recording booth so that the actors can see the animation. Second, you need some way to play the animation and record the actor's sound at the same time.

Unfortunately, no low-cost, off-the-shelf methods exist for doing ADR on a desktop studio system. You can certainly cobble a solution together using a sound recording device that is synchronized to a video playback machine. That's really all it takes, but it is a bit easier said than done. I am actively searching for inexpensive software that will run on a Mac or Windows system, and so far, this is all I can report to those of you who want to try to set up an ADR capability in house.

Sound Booth Monitor

The most inexpensive way of providing a monitor in the recording booth is to obtain a consumer TV that has a "video-in" jack on the back. Because the

Figure 3-18
PCTV Delux USB device provides TV support for your computer.

talent will only use the monitor to watch the lips of the character, the monitor doesn't even need to be color. Obviously, your computer must be similarly equipped with a "video-out" jack as well. You cannot connect the computer monitor output of a computer to most consumer television sets. If your computer does not have a video-out jack, you can buy a low-cost peripheral card for this purpose.

Pinnacle Systems offers the **PCTV Delux** card for $200 (see Figure 3-18). PCTV Deluxe is an external **Universal Serial Bus** (**USB**) device that lets you view, pause, and capture live video from your TV or any video source. PCTV Deluxe features TV stereo sound and hardware MPEG compression. The new high-speed **USB2** standard provides even higher quality and performance on your PC or laptop. The product comes with a 41-key PCTV remote control for all TV and digital VCR functions, so you can also kick back with a six pack and watch the game.

ADR Software

If you are already using Pro Tools, you can get a plug-in from Gallery Software called **ADR Studio** (www.demon.co.uk/gallery). This innovative software offers all the tools necessary for adding high-quality recordings onto a preexisting track (see Figure 3-19).

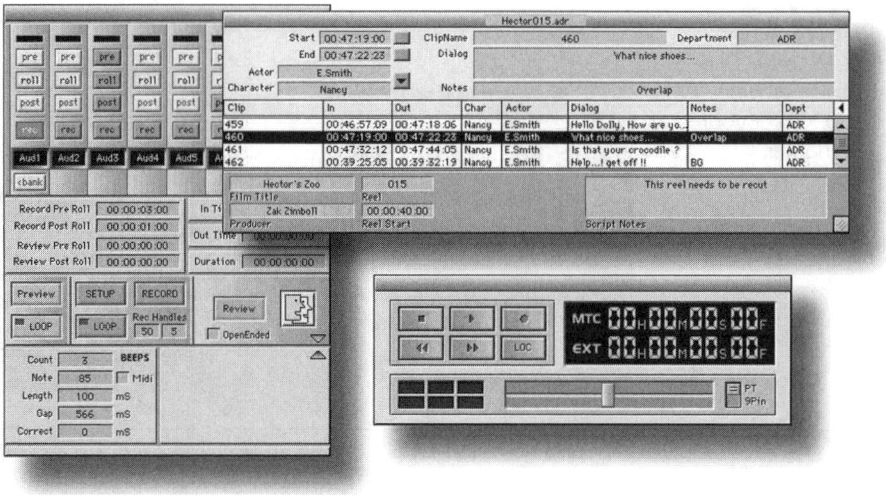

Figure 3-19
A montage of screens from ADR Studio, software that turns your Pro Tools system into a voice-looping studio. Very effective for post-synching animation.

Real-Life Story

Matthew Wood, the supervising sound editor for Skywalker Sound, needed a portable ADR unit for *Star Wars Episode I: The Phantom Menace*. Experienced in Digidesign's Pro Tools, Wood sought out a means whereby he could employ his Apple G3 PowerBook. "Magma was making a new expansion chassis," said Wood. "I got a Miro DC30 Plus and [a Digidesign] Audiomedia III card and was able to generate video off of the 8-gig internal drive and record the audio to a 4-gig expansion drive."

To get some extra help, Wood turned to Mark Gilbert, director of product support at Gallery Software. With Gilbert's help, ADR Studio software was installed into Wood's G3. The result was previewed for director George Lucas and worked flawlessly.

"Mark's ADR Studio drove the sessions," Wood says. "There's no way I could have done this without the ADR Studio." More information can be found at www.demon.co.uk/gallery/frameIDX.html.

Cabling

One should not discount the importance of the cabling that connects the various elements of your desktop system. Obtain the best cables you can afford and always treat a cable with great care. When budgeting for your studio, don't forget to set aside adequate funds for quality cables and connectors. The cables are your most essential component and are frequently the source of the most difficult problems.

If you are experienced with soldering and are meticulous in your work, you can buy both video and audio cable in bulk. You can then customize your own cables by soldering and crimping plugs and connectors as needed and solder the connections yourself. If you are not experienced, nor have the time, however, spend a bit more money and buy the ready-made cables from a reputable source. The trick is to plan your layout on paper and measure carefully.

The Role of Video in an Animation Studio

As a beginner in animation, you will most likely output your animation as some form of videotape. Later, as you progress, film output may become an option, but it is not likely that you will obtain film clients early in your career, and if you do, your input and output will most likely be in the form of large data files composed of large, still image sequences.

In most cases, your finished work will be delivered to your client in finished form, with or without sound, on a videotape. It is therefore necessary to have a VCR installed in your studio. This device is as useful in *playing* video as it is in *recording* video, so you might as well consider the possibility of using your VCR to *play* video *into* your animation system and using your computer as a digital video recorder. This is called **acquiring video assets**.

Acquiring video assets allows you to perform a vast array of special effects whereby you combine animation and video or directly animate the video. Once video is acquired, it resides on your hard drive as a sequence of distinct frames. These frames are usually stored in any one or more of the usual digital picture formats, known best by their acronyms:

- **mpeg (Moving Pictures Experts Group)** uses the file suffix **.mpg**.
- **tiff (Tagged Image File Format)** uses the file suffix **.tif**.

- **jpeg (Joint Photographic Experts Group)** uses the file suffix **.jpg**.
- **Targa** was developed by a company called **TrueVision** for their **Targa** series peripheral cards. It uses the file suffix **.tga**.

By accessing the individual frame files, you can manipulate them to create exciting adjustments to the real-life scenes that the file sequences depict. For instance, you can add an animated character to a scene of real-life actors or you can place real-life actors within an animated scene. This process is called **compositing**.

Video Acquisition

Although it would be possible to run a profitable animation business without translating your work to video or film in any way, it would be foolish to ignore the capabilities 3-D animation software offers for compositing.

In order for you to acquire video assets into your animation program, you will need a video acquisition peripheral card in your computer with a VCR connected to it.

Video Capture Cards

Similar to audio capture cards, video capture cards tend to fall into two grades: professional and consumer. Consumer boards feature one or more of the popular consumer video signal formats, such as **S-video, IEEE 1394 (Firewire)**, or a simple **RCA-type connector**[1] (the kind usually used as an audio connector on the back of a home VCR) for **composite video**. Although consumer video capture cards can certainly work for creating industrial-grade compositing projects, you should protect your upward growth by at least investigating the cost of a professional-grade video card. A professional card will offer both consumer signal formats as well as professional formats, such as Betacam (y, y-b, y-r) and/or **serial digital interface (SDI)**.

Other professional video card features include the capability to read **the Society of Motion Picture and Television Engineers (SMPTE) Time Code,** a signal added to professional videotape that enables compliant devices like a computer or VCR to identify every frame of a program.

---

[1]RCA [Radio Corporation of America] invented this simple form of audio plug.

SMPTE Time Code is an essential requirement in video editing and sound synchronization; it may be required for certain animation compositing applications.

For a **standard-definition** (**SD**) video capture, a card must be able to acquire 30 frames per second with a resolution of 720 pixels wide by 480 high (some cards use 486 as their standard). In the future, you may find that **high-definition** (**HD**) video comes into demand. It will require you to capture a wider range of frame sizes and rates, up to 60 frames per second, and 1,920 pixels wide by 1,080 high. Obviously, HD capabilities are only being featured in the highest level of professional capture cards.[2]

Wise computer manufacturers are beginning to catch on that consumers want to dabble in media production. Apple Computer seems to lead this trend and is packaging home and business systems with audio and video acquisition capabilities. By employing **Apple Final Cut Pro**, an NLE software product that enables acquired video to be edited, many amateurs are exploring the sophisticated realms of desktop professional video editing. If you were to obtain an Apple computer with Final Cut Pro and add a 3-D animation software package, you would have all the tools you needed for quality compositing.

Owners of Windows-based systems will be pleased to learn that a wide range of video acquisition cards are available, nearly all of which come bundled with NLE software.

One of the rude awakenings experienced by interns and beginning animators who work in my shop is the fact that 3-D animation software is not equipped to play the finished animations you create. Although the software can create rather detailed previews based on the capabilities of the host computer and video card, these previews are not, by a long shot, the finished work.

In order to see your finished work in all its glory, you must assemble all your picture files into a sequence and then play that sequence at full resolution (most commonly, 720×486 for SD and $1,080 \times 1,920$ for HD) in real time.

Such capabilities were long reserved for NLE software, often equipped with specialized peripheral video boards. In the early stages of building an animation business, you might have taken your sequences to an outsourced

---

[2] As of this writing, JVC has introduced a new format of HD, recorded on standard mini-DV tape. The first component of this system is the DY-HD10 HD Camcorder, introduced at the unheard price of $3,995.00. A standard IEEE 1394 (firewire) card is sufficient to capture this camera's output to a Windows XP computer for subsequent nonlinear editing.

facility, such as a video producer or video-editing house. At the editing house, your frame sequences would be edited frame by frame into a final, playable scene. Such labor!

Today, faster, better, less expensive solutions are available. To save you the hassle of converting your frames into a playable, compressed format, you can use a flipbook application such as **IRIDAS FrameCycler**. A full version of FrameCycler is included on the CD that comes with this book. All you need to do is get your complimentary license from IRIDAS (www. iridas.com) to get started right away.

FrameCycler enables you to load a frame sequence into RAM and play it back at full speed. It even lets you use a sound file, and it can handle multiple alpha channel layers for quick compositing tests. In addition to playing SD sequences, FrameCycler can play HD and even film-resolution (2K) frame sequences. This is a useful tool that can save you thousands of dollars in hardware or facility rental charges.

Another method for viewing your sequences in real time is to install an NLE system in your computer. Currently, two types of NLEs exist. The first type depends on installing *proprietary boards*, so I call them **PB NLEs**. The second requires no boards but relies instead on the computer's CPU to process all the editing functions. I call these systems *software-only* (**SO**) **NLEs**.

Obviously, the least expensive are the SO NLEs, which sell for as little as $79 (the **Studio8** NLE from **Pinnacle Systems**.) Because these NLEs still require a video acquisition card for importing and exporting video to and from your computer, you should not assume that the software is all you need. However, many of these programs come with an inexpensive video acquisition card, which is adequate for your needs as long as you are working in the digital video environment. If you are working in **composite video**, **component video**, or with a VCR employing an **SDI** capability, you will require more expensive boards and the choice of a PB NLE may be more logical.

If and when you choose to install your own NLE software, you will no doubt find your solution among one of the five leaders in this field: **Apple, Matrox**, **Pinnacle**, **Media 100**, and **Leitch-DPS**. Each company is a leader in its field because it manufactures *both* the hardware and designs the software for its brand. Companies that supply both hardware and software are more reliable, because they represent single-point responsibility for any problems that might develop.

When choosing NLE software, it is wise to match the system's capabilities, particularly the types of sequence files it uses, with the output of your animation software and the needs of your client. Apple's **Final Cut Pro**, as of this writing, will only work with digital video, making it an unwise solution if your clients require nondigital-video masters.

Leitch-DPS, for example, remains the only NLE manufacturer with products that can convert and store all popular file formats as soon as your software creates the sequence. This firm's **Reality** package (which comes with their proprietary board set) has been my favorite tool for rendering finished animation sequences since 1992 when DPS invented the first economical desktop **digital video disk recorder** (**DDR**). DPS products all work in real time with a direct-to-NTSC signal and therefore are a bit more costly than the average video acquisition/playback card that requires CPU rendering to produce a recordable signal.

Pinnacle Systems, on the other hand, offers a wide assortment of products, including both hardware and NLE solutions beginning with the Studio 8 and **Edition DV500**, which comes with a digital video capture/playback card and real-time playback to an NTSC monitor. The professional features found in Edition rival those found in Final Cut Pro and other systems costing quite a bit more. These low-cost systems represent Pinnacle's consumer division. Pinnacle's professional division is more expensive, offers more features, and provides a higher quality, but for beginning animators, the consumer products are all you really need for now.

The VCR

Currently, video is recorded on videotape (although video decks and cameras with removable hard drives are now being offered). Many types of videotape, recorders, and cameras are available, and each may be grouped in terms of quality.

Historically, videotape began as an **analog** form of recording. Analog recording captures picture information in the form of continuous waves of electrical force. Many forms of analog tape are still in use today, but the most current method of video recording is **digital**. Digital video recording captures picture information in the form of just two numbers: one and zero. Many types of analog and digital recording are available on the market today, and your clients may require more than one. Currently, the most common formats for delivering animation are BetacamSP (a discontinued analog format invented by Sony), DVCam, and MiniDV.

Eventually, digital formats will prevail since they are superior to analog formats. Digital recordings encompass less noise than analog, may be duplicated without loss from copy to copy and allow for more precise control of the qualitative elements that comprise the picture.

Each VCR offers one or more modes of transferring video signals to a computer or another VCR. In order of increasing levels of quality, these modes are **Composite**, **S-Video**, **Component**, **IEEE 1394** (**Firewire**), and **SDI**. In evaluating your choice of VCR, you should strive to obtain the highest-quality transfer mode possible. Obviously, the mode of the VCR should match that of the computer's video capture card. Some VCRs and video capture cards offer additional modes as optional plug-in boards, enabling an upgrade after the initial purchase.

When choosing a VCR, you must anticipate the types of tape your clients may send you and subsequently want to receive as masters. This decision will also be based on your budget. Obviously, if a single VCR can handle several formats at one time, you will have a greater chance of having the right one in house when clients come knocking.

Currently, a typical multiformat digital VCR offered at a reasonable price is the **JVC BR-DV3000U** (see Figure 3-20). It plays back and records on **MiniDV** and **DVCam** (large and small cassettes without an adapter).

Of course, if you'd like maximum flexibility at short-term cash savings, don't buy a VCR at all—rent one. Daily rental rates for VCRs are as little as $150, depending on the model. If you rent the deck on a Friday afternoon and return it Monday morning, you can sometimes get the entire weekend for the same price as a one-day rental. Rent for four days, you can usually keep the deck for a week. Keep in mind, however, that you have to be near enough to a rental facility to make this option

Figure 3-20
The JVC BR-DV3000U can use a wide range of digital video cassettes and sells for $1,995.

*practical. If you are renting the same deck for more than a month, it
may make better economical sense to buy it. Also consider buying a good
used VCR to save as much as 50 percent off the list price.*

Video Monitor

In addition to a VCR, you may require a few monitoring tools to evaluate
your video quality. Although you can easily display the video output on your
computer monitor, you may want to add a video monitor for a direct view of
the VCR's output. Video on a computer screen is not the same as video on a
TV screen. The video monitor can assist you in determining the source of a
problem, should the video not appear as you anticipate. With a video monitor, you can use the process of elimination to determine the source of a
problem.

Video monitors are available in a wide range of qualities and costs, with
the least expensive running a few hundred dollars and the most expensive
—one suitable for network applications—costing several thousand dollars.
Avoid using a consumer TV set for this purpose. A consumer TV does not
offer professional-quality image controls and inputs. Professional monitors
distinguish themselves by offering higher-quality components, professional
cable connectors, signal analysis tools, and a wider range of adjustments
than consumer sets. Make sure that the monitor you buy has the same signal mode inputs as your computer and VCR.

Summary

You have now been exposed to the various hardware configurations that
you will employ to build your own 3-D animation studio. The object is to get
you started with an affordable package that can be quickly amortized by
professional assignments suited to the package you can afford. Once you
gain experience and clients, you can expand your facility to include a wider
assortment of services or build multiple workstations to execute more work.
Now let's begin our exploration of the leading 3-D animation software
products.

CHAPTER 4

The Mechanics of Three Leading Animation Programs

Three Leading Animation Programs

This chapter introduces you to the three animation programs that serve as the basis for this book. One of the ongoing debates in computer animation is which program is the most competitive, which the fastest, or which the best. This is a difficult question to answer.

Which Is Best?

First, because of the highly competitive nature of the industry, each program is, at any given moment, under rigorous development. Maya today might be a totally different program when this book hits the stores, and different again six months later. In other words, we're talking about moving targets. Each program is morphing as we try to hunt it down and define its qualities. However, one can distinguish a certain personality in each of the programs.

Second, no clear-cut statement can be made as to what is the absolute best animation program. What might be best for a character animator may not be best for a logo animator or a game animator. However, a trend in the leading programs seems to be occurring at present and is related to two qualities: *ease of learning* and *depth*. Ideally, a program should be both easy to learn and deep in capabilities. Unfortunately, none of the programs I have seen demonstrate both qualities. Consequently, you can at present build a contrary scale of these two parameters.

I encourage you, as you read this chapter, to start exploring the CD that came with this book and using the various aspects of each program. You will soon acquire a personal sense of which program feels most comfortable and suits your needs. That is why the CD is there.

In my opinion (okay, start those letters and emails flowing), I would classify New Tech's **LightWave** as the easiest 3-D animation program to learn, although it may be shallow in terms of some of its facilities, especially where character animation is concerned. Discreet's **3ds max** is more difficult to learn than LightWave, but its more robust features increase its depth. Both **Softimage XSI** from Avid and Alias|Wavefront's **Maya** are at the top of the difficulty spectrum. These programs are difficult to learn out of the box and are deep in facilities. For example, I have not yet met an animator proficient in XSI or Maya who did not attend a school to learn it.

Why can't a program be both easy to learn and deep? Could it be that the deep program manufacturers, while interviewing users in focus groups, have identified difficulty as a *positive* attribute? Hanging out with some Maya and SXI animators, there definitely seems to be a kind of fraternal atmosphere in the air, perhaps bred from hazing and tortuous *rites de passage*.

Third, the real measure of one program's superiority over another could be defined by which program generates the most revenue. This would be a difficult statistic to obtain, but a rough review of major films', games', and network televisions' uses of computer animation over the past five years would probably result in a pretty even spread among the major animation programs. At certain times, one program might reign in one category or another, but as each produces new features and gathers new operators under its umbrella, different programs take their turn at the head of the class. None, however, as yet totally dominates, which is a good thing.

This chapter will dissect the basic mechanics of each program. Mechanics can be broken down into the following aspects:

- **Views**   How you see your work
- **Objects**   How you make and manipulate objects
- **Timeline**   How you add the fourth dimension to create animation

Common Elements

As the technology and art of 3-D animation has progressed over the past decade or so, certain conventions have come into existence and are common to most, if not all, animation software. Here are some of the conventions shared by LightWave, 3ds max, and Maya. Further on in this chapter, each software program will be discussed separately to give you a bit more detail.

By the way, let's define the word *object* here so nobody gets confused. For the purpose of this book, object can refer to any point, collection of points, polygon, collection of polygons, surfaces, or effects. That sounds like the contract I signed to write this book! The point I want to make is that objects can be an orange, a wall, an ocean, a streak of light, a puff of smoke, or even an infinitesimal point that cannot be seen anywhere in the animation (called a *null object*). Cameras and lights, by the way, sometimes act like objects too, because they can be animated in much the same way as an object. I try, however, to distinguish cameras and lights whenever it makes a difference, such as **deformation**. You can't deform a camera like you can deform an object, at least not in the virtual world.

Views

Each program offers four methods for viewing a scene, three of which are **orthographic:** the front view of the X- and Y-axes, the top view of the X- and Z-axes, and the side view of the Y- and Z-axes. The fourth is a **perspective** or pictorial view. Orthographic views are devoid of perspective and intended to be simple, clinical, blueprint kinds of displays based on a 3-D Cartesian plane grid. Orthographic views are useful for creating, finding, and changing specific elements of a design, such as a seam or points, and for accurately laying out animation.

The perspecive or pictorial view allows the scene to be seen as the human eye might see it, principally from an external perspective, with the ability to rotate the object(s) or the point of view around all three axes. The perspective view is usually offered from some omniscient point of view through the software's camera(s) or even through a lighting device. At any time, the operator may switch to one view or any combination of views, depending on the need.

Program Navigation

Animation programs are so robust they cannot possibly display all their tools and features in one screen. Consequently, some method of navigation is featured in each program. The most common method is to feature a range of **category buttons** and menus and **subcategory buttons** and menus that represent ever more specific subsets of commands. For instance, a category button might be Create, and a subcategory button might be Primitives, and a further sub-subcategory might be Sphere. Category buttons remain onscreen at all times. When one is clicked, an appropriate subcategory of buttons may appear. This is usually referred to as "drilling down" into a program.

Other buttons may also remain onscreen at all times but are not category buttons. These pertain to functions that must always be available to the operator, such as **File** and **Save As**. We could consider these to be **constant buttons**.

Category buttons and subcategory buttons may be simply **alphanumeric** (such as those in LightWave), **iconic** (such as many in 3ds max and Maya), or a combination of the two. Learning the various pathways of the program and getting to various tools when you *know* the program eats up a lot of operator time. As a result, program developers are always trying to find new and easier ways of learning and operating the navigation. More details are provided on each program later in the chapter.

View Navigation

View navigation refers to the method each program provides to manipulate what the views are shown. The navigation is usually done by the mouse or pointing device, often in combination with a keystroke. Although some similarities exist, each program is different. Obviously, orthogonal views cannot allow tumbles (because the tumble would redefine the view's definition), but "on-axis" adjustments are possible to adjust the view. Keep in mind that on-axis moves are not camera movements (such as tilts, pans, or zooms) that imply a pivot point in the camera head. Perspective views, however, are intended to allow the operator to examine the scene from any angle or distance, so full movement, including tumbling and zooming of the view, is supported.

Object Navigation

The tool set for affecting objects is called the **manipulators** (not to be confused with your bosses). Today most 3-D animation programs have standardized the onscreen symbols that users employ to execute object manipulation. Using symolic manipulators, animators may manipulate cameras, lights, bones, and objects. Object maniulation is the basis of animation.

Although object navigation allows any object in the view to be selected and manipulated in many different ways, object manipulation allows a selected object to be changed. Object manipulation is divided into two categories: *movement* and *deformation*. Movement includes changing the object's position on the stage and rotating the object. Deformation includes any change of the object's size, shape, or surface characteristics.

The most common manipulators are **move**, **scale**, and **rotate**. When any of the manipulators are selected, an appropriate symbol appears within the selected object or element. The move symbol is usually composed of a box outline with three colored vector lines (usually with arrowheads at the ends) representing the three axes (red X, green Y, and blue Z) emanating from it. You may move the selected object freely on all axes if you click the central symbol, or you can isolate one axis by clicking one arrowhead at the end of that axis.

The scale symbol is similar to move, but the center symbol and the ends of the vector lines will change to indicate a different manipulation from move. The colors of the vector lines and end symbols are the same as those for the move tool, and they refer to the same axes as the move tool. When the center symbol is clicked, the mouse affects all three axes simultaneously.

The rotate symbol is composed of three intersecting rings (red for the x-axis, green for the y-axis, and blue for the z-axis) around a central point that represents the pivot point of the object. Again, clicking the center of the symbol allows free manipulation of all axes at once, whereas clicking any one of the colored rings selects a single axis. It should be noted that Light-Wave tends to vary from this convention somewhat, as described later.

Animation programs enable the operator to obtain a view through the virtual camera and, in some programs, through a lighting instrument. Object navigation and manipulatins, unlike view navigation, may be recorded as keyframes at specific locations on the timeline and provide a powerful animation element.

Layers and Histories

Animation, particularly the modeling phase, is a kind of **iterative** process. This means that you tend to perform a step, evaluate it, and then reiterate the process to a finer level of detail and/or satisfaction. When satisfied, the animator will proceed to the next task in a methodical progression. In the process of modeling an object, you might execute hundreds, perhaps thousands, of iterations. Most programs have tools for assisting this meticulious process.

The program can assist you by automatically recording each of your decisions in a list, often called a **history**. By opening this list, you can go back over many decisions and find a place where perhaps you went off on an unproductive tangent. From that point you can begin again without redoing a lot of work.

Layers are another characteristic common to all three programs and are useful in building objects. For instance, if you were building an apple, you might create half the circle in one layer, duplicate it to another layer, and mirror it to form the circular shape. In another layer, you might design the stem and in yet another layer, a leaf for the stem. Each layer enables other layers to be superimposed so the various elements can be aligned with one another. At some point, you may save all the layers separately to form a file collection that represents the entire object, or you may combine any two or more layers to combine all the elements together into one object. Each program has different features to its layer module that are explained in more detail later.

Timeline

Most 3-D creation programs feature timelines as a way of defining and creating animation. The timeline enables you to initiate and control the animation, specifically the manipulation of objects, cameras, and lights *during set lengths of time*. Animation created on a timeline starts with frame zero. Any object that appears in the animation, whether it changes or not during the animation, must have a keyframe on the timeline. The same goes for the camera(s) and any lights.

A **keyframe** is usually represented as some small mark on the timeline, but its purpose is massive. Within the keyframe, all the attributes of the object are stored. If none of these attributes change during the length of the timeline, nothing will change in the animation.

In order to create animation, the operator must add keyframes for any object, camera, or light that is to change during the duration of the timeline. Each element that will be animated is selected by the animator and *effected* or changed, such as a change of position or size. Once that change meets the animator's satisfaction, a keyframe is created at some point on the timeline.

If the object's attributes continue to change, another keyframe for that object must be entered on the timeline to record the changed attributes. The computer then calculates automatically the changes from one keyframe to another in a process that is known, quite aptly, as **in-betweening**. In other words, the frames *in between* any two keyframes are calculated and drawn by the computer. This is the main difference between computer and traditional hand-drawn animation.

Most objects are introduced to the timeline at frame zero, but it is possible to introduce objects on any frame. Such objects may be introduced on camera by popping, fading, or being brought into the scene from off camera.

Let's imagine, for instance, the animation of a paddleball on a rubber band, the kind of thing that frustrated the heck out of me at a carnival decades ago. The zero keyframe might define the starting position of the ball as it zooms off into the air away from the paddle. The second keyframe would be where the rubber band tightens up enough to start pulling the ball back to the paddle, where it will inevitably miss and whack my eyeball out. The third keyframe might be my doomed eyeball. If we were to watch this animation as executed by an animator with as much programming skill as the paddleball skill I demonstrated for my girlfriend at the Mercer County Fair in 1966, the result would look really artificial. The ball would zoom out, pause a brief instant, and then zoom back, all at the same constant speed.

In reality, as you well know, the ball zooms out but starts to slow down in infinitesimal degrees until it kind of hovers for a moment in space. (This is the precise point, by the way, where the ball figures out where your paddle is and calculates the exact trajectory to your eye.) It then begins to accelerate back to its inevitably ophthalmic destiny. In order to create such an effect in a 3-D animation program, you must impart extra information into the computer. This extra information relating to subtle variations in the in-betweening is usually stored in one or more keyframes. As you might imagine, this information is difficult to visualize. Computer programs usually represent these subtle changes as graphs with curved lines. The names for these editing tools vary from program to program.

In traditional cell animation, subtle acceleration and deceleration effects are created by gathering more images of the ball at the end of the rubber band's tension, thus making the ball appear to slow down. You would then place fewer images of the ball where the speed is greater. A falling ball might look something like the orange in Figure 4-1.

Notice in this 3ds max screen view the gathering of wireframe images as the orange nears the impact point. Also notice in the timeline the grouping of keyframes around the impact point (frame 20). Notice also that between frame 19 and 20, the orange is vertically squashed, and between frame 20 and 21, the squash is reversed.

Now let's examine each of the program's unique features with respect to the general aspects defined earlier.

LightWave

LightWave was introduced by NewTek as a module within the *Video Toaster* (a comprehensive video production tool) in 1991, and quickly developed an independent following among a growing number of new animators.

General Description

LightWave is actually three separate programs in one. One program, Modeler, handles all the modeling functions. Another program, Layout, is where all the scene-building and animation functions occur. Within Layout, the

Figure 4-1
In traditional animation, an animator would create more drawings and move the object fewer increments whenever he or she wanted the orange to slow down or speed up.

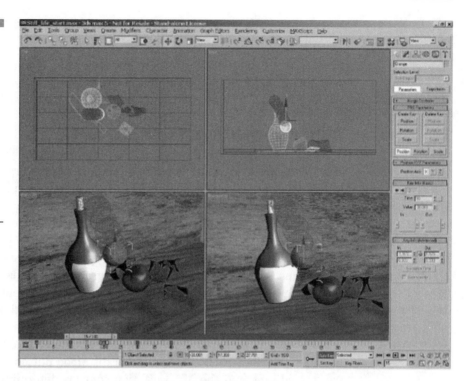

third program, ScreamerNet, LightWave's renderer, resides. It should be noted that not all 3-D animation programs include their own render software, and ScreamerNet is excellent. An animator can have both Modeler and Layout open at the same time, even while rendering on ScreamerNet, switching as needed between all three. Of course, the animator may choose to save RAM space by having only one portion of LightWave open at a time.

LightWave can create animation sequences and still frames that range from 16×16 pixels for use in web site GIF animations to 8,000-pixel-square frames suitable for cinematic projection. **D1** and **D2** (NTSC and PAL) professional formats are among the many preset defaults, as is AVI, for Internet use.

A typical workflow in LightWave would entail the following steps:

1. Acquiring or using Modeler to create objects and storing them in appropriate categories

2. Loading objects into Layout

3. Assigning surfaces and textures to objects

4. Adding and orchestrating lights within the scene

5. Animating the scene by generating keyframes for objects, lights, and/or the camera

6. Assigning resolution and camera attributes

7. Rendering and saving image sequences

Program Navigation

The Modeler and Layout screens are similar, offering a large, central display area. Above and below the display area are buttons particular to either Modeler or Layout functions. The most often used buttons are the **panel tabs**. These are category buttons designed to appear as file tabs in the left side of the top row. These select the major categories of action in both modules.

A narrow row of buttons along the left edge of the screen is divided into two groups, upper and lower. The upper group of buttons contains menus that deal with file activities, displays, and an assortment of editors. The lower and largest subcategory of buttons opens menus that change as the panel tab buttons in the top row are switched.

In Layout, these subcategory buttons are grouped into Edit Mode selections and Mouse Action controls. The Edit Mode buttons determine what you are editing, and the Mouse Action controls determine what happens when you slide the mouse.

In Figure 4-2, you can see a typical Modeler screen set up with four quadrant views (three orthogonal: top, back, and right) and one perspective. In this figure, titles and arrows indicate the various controls of the interface.

Figure 4-3 is a typical LightWave Layout screen, with the default display settings (**Layout** > **Options** > **Display options**) set to **OpenGL shading**. Even on a laptop (this frame was grabbed from an IBM A31p that was simultaneously running Modeler, 3ds max, Microsoft Word, and Explorer), the OpenGL functions of the video card display realistic shading with little lag time.

Notice the timeline displayed below the views. The play cursor is paused at frame 28 as the rotation of the airplane is adjusted. This causes a keyframe to be automatically added to the timeline and is shown as a vertical yellow line on frame 28. Other keyframes at frames 0, 25, and 30 define the path attributes of the fighter, which include position and rotation. The keyframes displayed on the timeline are only those for the **Current Item** (m_body11) shown in the collection of **constant buttons** below the timeline. Other items in the scene could be keyframed by using the pulldown menu. If another item were selected, the m_body11's keyframes would be replaced by the keyframes associated with the new item.

Figure 4-2
The Modeler interface screen. Notice that all the buttons in LightWave are alphanumerically labeled with keyboard equivalents on each button. No picture icon buttons are used in LightWave.

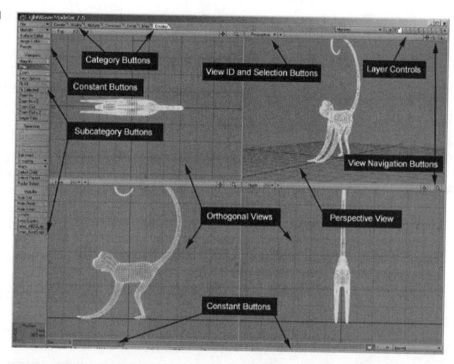

Figure 4-3
The layout interface in LightWave. This is one of 11 default divisions of the working screen (perspective, top, and right views).

Similarly, by using the buttons below the item selection tool (Objects, Bones, Lights, and Cameras), different classifications of items can be selected and applied to the timeline. The capability of LightWave to segregate categories of items and individual items keeps the timeline uncluttered.

To the right of the item selection buttons are the **keyframe buttons** that enable the creation and deletion of keyframes. Further to the right are the buttons that control the timeline's playback, similar to those on a VCR. See Figure 4-3.

The typical work routine involves starting a LightWave project in Modeler, where the animator creates, modifies, and/or acquires all objects needed in the project. The objects are then stored (the default suffix is .lwo) in an Objects subdirectory, which should be nested within the Project subdirectory. If the objects are created or stored at the **origin** (where x, y, and z equal zero) on the Cartesian plane, they will be retrieved in the same place in Modeler.

Many common real-world objects consist of complex subassemblies of component objects. A hair dryer, for instance, might have a handle, a fan, a cone, switches, a cord, and so on. These subassemblies should all be stored as a set with a common reference to each other, preferably at the origin. Consequently, if you are designing a room full of objects, it might save time to save each object in an arrangement around the origin so that you need not move them around much in Layout. It's your choice.

In Modeler, the animator collects all the objects for a scene and stores the scene as a scene file (the default suffix is .lws) in a Scenes subdirectory within the appropriate project directory. Animators usually create static scene objects first, then the objects that have movement, and lastly any required characters. When all the modeling is completed (or far enough developed that someone can start animating), the objects are imported into Layout where animation begins. LightWave Modeler and Layout are closely linked so that any changes to a model in Modeler can be easily updated in whatever scenes are opened in Layout.

View Navigation[1]

Ten views are available in LightWave: Top (XZ), Bottom (XZ), Back (XY), Front (XY), Left (ZY), Right (ZY), Perspective, Light, Camera, and Sche-

---

[1]All descriptions are in terms of Windows keys and routines. For the Macintosh equivalents, please refer to each program's individual documentation.

Figure 4-4
The LightWave Layout screen divided horizontally between the schematic view and the camera view

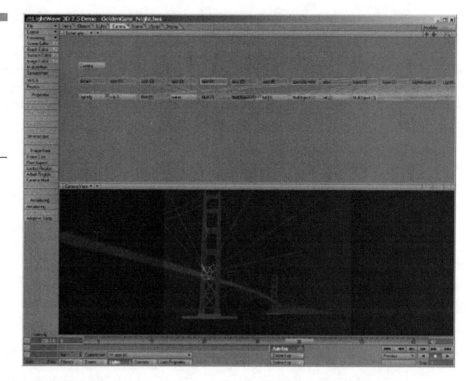

matic. They can be selected by keystrokes or pull-down menus, and be grouped on the screen in any combination and size.

All 2-D views in LightWave are targeted to the same **aimpoint** in 3-D space and are linked to one another. Adjusting one 2-D view will cause appropriate adjustments to all other 2-D views.

The 3-D perspective view in LightWave can be manipulated by pressing the Alt key and moving the mouse. The view can also be changed by clicking and holding any of the navigational buttons (see Figure 4-4) in the upper-right corner of the screen while moving the mouse accordingly. The 3-D light and camera views can only be moved when their positions are changed. These views facilitate targeting the lights and camera at specific objects in the scene.

Each window in LightWave can be set up to display its contents in a variety of shading methods, including wireframe, color wireframe, sketch (a combination of wireframe and flat shaded), flat shade, smooth shade, weight shade, and texture.

The **schematic view** displays all the objects, the lights, and the camera within the scene as a block diagram or **hierarchical display**. The schematic view can display three or more rows of blocks: one for the camera, one for lights, and one for objects. Lines, linking the blocks, represent relationships between elements, such as targets or parent-child links.

The animator can click any block to select an element for attention. This display is useful for finding and selecting a specific element from a complex scene where that element might otherwise be difficult to identify. A large spaceship model, for instance, might have a lever on a panel in a control room within the ship that would be extremely difficult to locate without the schematic view.

In Figure 4-4, notice how a light "spot (4)" has been selected in the schematic view and how in the camera view the light is displayed with its light pattern defined as dotted lines. In the constant buttons at the bottom of the screen, the category "Lights" is illuminated. Above that button, the actual name of the light, "spot (4)," is indicated as the current item. The schematic method enables a simpler display of active elements in the camera view while providing the animator with an easy method of element selection.

Object Manipulation

Objects that can be manipulated in LightWave include 3-D objects and elements, bones, lights, and the camera (only one camera object is permitted in any LightWave scene).

Object navigation tools are found under the Items tab in LightWave. These include Move, Rotate, Size, Stretch, Squash, and Numeric, all of which can be selected by clicking on named buttons when the Items menu is displayed (see Figure 4-5). With other menus displayed, the tools can be accessed by keystrokes (shortcut keys are printed on all the buttons in LightWave to make memorization easier).

When activated, the conventional colored symbol relating to the type of tool selected is superimposed on the center of the object. Three colored lines with arrows represent moves, intersecting circles represent rotation, and three colored lines with boxes at the ends represent squashes and stretches. The pivot point of the object (the point around which the object rotates) may also be moved (using the Move Pivot Point command). A reset command also restores all current navigations since a keyframe was last created or set.

LightWave differs somewhat from the object manipulation convention established by other animation programs. Originally, LightWave enabled

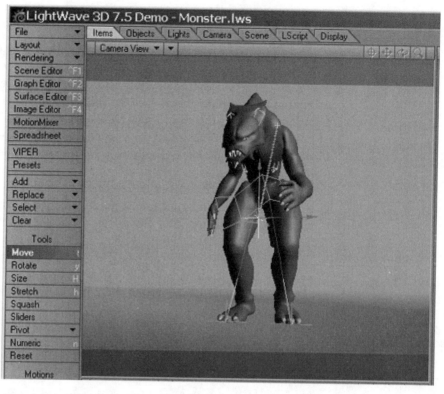

Figure 4-5
The object navigation buttons under the Items category in LightWave. Notice the movement indicators (red, green, and blue arrows) on the monster.

the user to select axes by choosing different mouse buttons, rather than clicking the colored symbols onscreen. Now both methods are supported, but LightWave does not support the convention of selecting the center point of each symbol to allow multiaxis manipulation, because this would negate LightWave's original mouse-button-specific convention.

When a navigation button, such as Move is activated, the mouse's right and left buttons control the object when the cursor is dragged anywhere onscreen. Of course, the appropriate colored navigation tools (much the same as 3ds max and Maya) can also be selected, but this seems unnecessarily tedious when the mouse buttons can be used so generically.

Selecting the Numeric tool opens a window that enables the user to manually type in numeric values for each parameter. This is useful for obtaining more precise results than can be obtained with a mouse, and this maintains consistency when working with several matching objects that must be effected identically.

Object Creation

You'll learn more about the theory of object and surface creation as a general topic in Chapter 5, "Objects and Surfaces," and by working the tutorials on the free CD that comes with this book. In order to give you the navigational highlights, this section discusses how you make objects in LightWave.

Points, curves, polygons, and solid objects are created in LightWave's Modeler module under the Create tab. In the vertical button bar, you will find a wide assortment of primitives from a simple box to a helix, a gear, or a toroid. Using the primitives in just one plane, simple polygons (such as circles, rectangles, and so on) can be made. Below the primitives, you'll find a button to make points. The *Make Pol* button, for example, joins selected points into polygons (see Figure 4-6).

Spline and Bezier curves can also be created and then extended in three dimensions to form shapes. In LightWave, which is a polygon-driven program, the spline and Bezier forms should be converted into polygons prior to rendering.

Under the Modify tab, you will find tools to change the objects you made under the Create tab. These tools enable you to form highly irregular shapes that can be extruded into solids or cut and pasted into complex forms.

When modeling from hand-drawn sketches, it helps the artist to input the drawing to the animation program for use as a template. Scanned images can be imported into LightWave and placed in any of the orthogonal views for reference during modeling.

Exercise

Here is an exercise that gives you a chance to put these concepts into practice. First, launch LightWave Modeler. You should already see a four-view workspace and the active tab should be Create. In the left column, under Objects, click to select Ball. Move the cursor to the origin (where x = 0, z = 0, or the center of the crosshairs), and click and hold. Drag out a circle. If you mess it up, go to the Undo button at the bottom center of the screen and click it.

Move your cursor to the Back view. Click and drag anywhere in the view. See how your mouse movements create and stretch the third dimension of the sphere. Release the mouse, and click and drag elsewhere. See how clicking in various areas of the view changes the effect on the sphere.

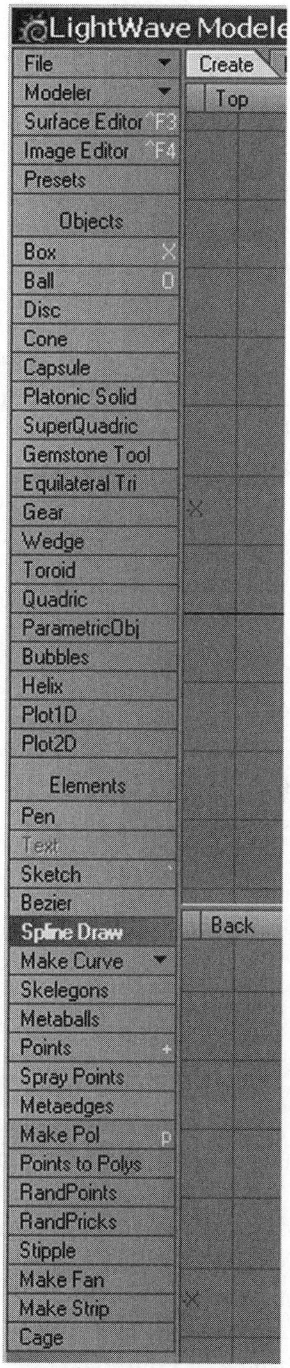

Figure 4-6
LightWave's Modeler Module under the create tab

Repeat the click-and-drag action in any of the other views. Finally, adjust the object so it looks as perfectly spherical as you care to make it.

Object Editing

Object editing begins by selecting all or part of an object and then performing the editorial function. LightWave Modeler offers three modes of selecting an object: *points*, *polygons*, and *volume*, all of which are represented as constant buttons in the lower right of the screen. When selecting points or polygons, the animator clicks and holds the mouse to select one or more targets that become highlighted in yellow. Releasing the mouse changes the cursor into an eraser. Further clicking a selected choice unselects it. Holding the Shift key enables the cursor to continue to select targets.

Incidentally, selection only works in views that are set to display wireframes. Don't quibble. It's too hard to select a point on a smooth, shaded surface anyway.

Using the Volume mode, a blue box is extended in all views simultaneously. In the Include volume mode, anything in the box is selected. In the Exclude volume, everything outside the box is selected. The spacebar can cycle through the selection modes.

Once selected, object elements can be affected by selecting the appropriate buttons in the left vertical column or by using cut, paste, undo, and redo buttons in the lower-right corner of the screen.

Layers

In Modeler, LightWave offers 10 layers for editing objects, and their controls are in the upper-left corner of the screen. Each layer is represented by a rectangle divided by a diagonal line. Clicking the top half of any layer rectangle selects that layer as active. When a layer (or layers) is active, the contents of that layer are displayed in a bright color (wireframes, for instance, are bright white when displayed in an active layer), and they can be edited.

Clicking the lower half of the layer rectangle makes that layer (or layers) a background display layer. Anything in that layer is displayed in all the views as a darker contrast color (black in the wireframe mode). The judicious use of background layers provides a means of accurately orienting the active layer for comparative editing (see Figure 4-7).

Figure 4-7
The perspective view in Modeler, showing the second layer active and the first layer as background

In Figure 4-7, a portion of the monkey's tail was selected using the volume select mode (indicated by the blue box) and was cut while layer 1 was active. Then I switched to layer 2 and pasted the tail there, where it became white, indicating it was in the active layer. Finally, I clicked layer 1 into background mode, which reveals the rest of the monkey in black for a reference to any further work I do in layer 2. Notice the layer buttons at the top right that show layer 2 as active and layer 1 as background.

Exercise

Here's another exercise. Take the sphere you made in the previous exercise or remake it if you already kissed it goodbye. At the bottom of the screen, click the Volume button. This tells the program you want to select a 3-D volume. In any view, click and drag your mouse to contain slightly more than half the sphere in the faint blue box that appears when you click and drag (this is the volume you are selecting). Use one or more views to adjust the size of the volume. At the bottom of the screen, to the right of the Volume button, is the Cut button. Click it once, and more than half the sphere vanishes.

In the upper-right corner of the screen is the layer control. The leftmost square has a small black triangle within a larger triangle. This indicates that the first layer has an object or objects in it. Click on the next layer to the right to select the second layer. The rest of the sphere vanishes because you are no longer looking at the contents of layer 1. In the bottom of the screen, click the Paste button. You have now pasted the cut portion of the sphere into the second layer. A small black triangle now appears in the second layer box.

Click on the lower half of the first layer box. You'll notice a black "ghost" image of the other portion of the sphere (in layer one) appearing as a reference in layer two. In the tabs, click Modify, and in the left column, click Move. In any view, move the half sphere wherever you like (notice you can't move the ghost image of the portion in layer one.)

Now click Cut, select layer 1 in the layer controls at the top left, and click Paste.

You've now rejoined the cut portion back to the original layer, but after modifying it in terms of position. Try this exercise with other portions using other buttons in the Modify tab to become familiar with the cut and paste power of LightWave layers.

Timeline

Two timelines are used in LightWave's Layout module (for obvious reasons, none is shown in the Modeler module). One of the timelines appears across the bottom of the Layout screen at all times. Its default length is 60 frames, but typing a different number in the input window to the far right of the timeline may change this. This timeline is principally for an immediate reference at all times, and as mentioned earlier, it displays the keyframes for the current item selected.

A much more detailed timeline exists under the **Scene Editor** panel (a constant button at the top of the left column). Here every object in the scene is shown as a separate horizontal band, similar to the many layers of a **nonlinear editing** (**NLE**) program. The timeline automatically lengthens itself horizontally to the position of the latest keyframe created for any object. It also expands vertically with every object added to the scene.

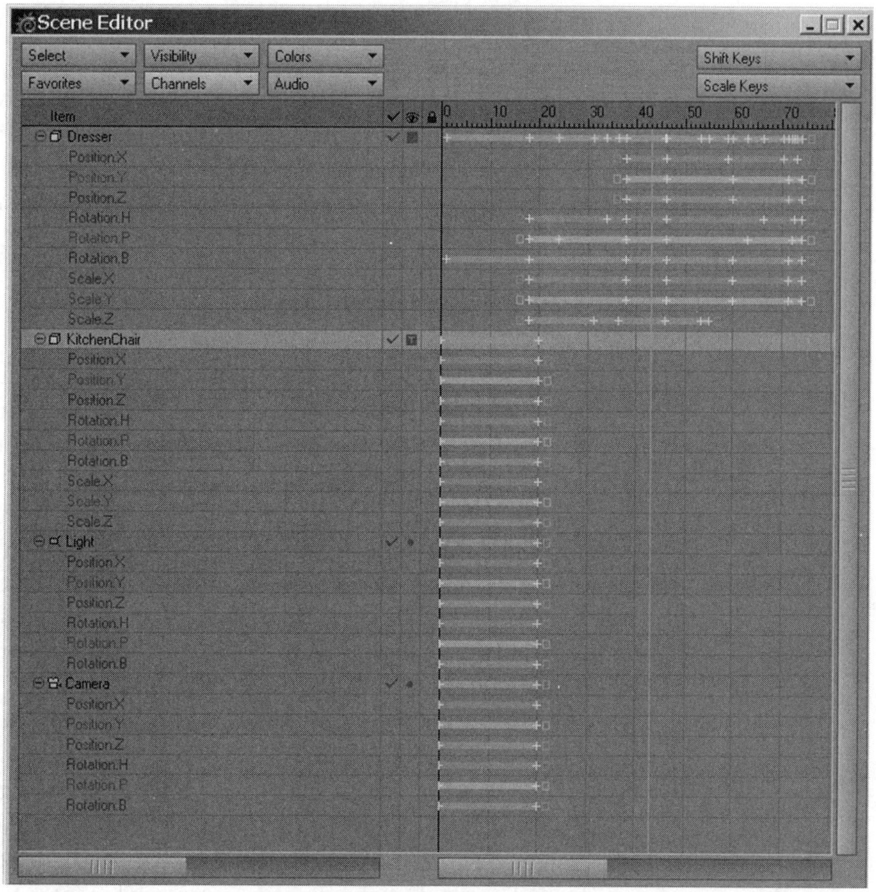

Figure 4-8
The most elaborate timeline in LightWave is the Scene Editor, where every channel can be examined and adjusted by mouse drags.

Each ribbon on the default Scene Editor display is actually a nested collection of ribbons called **channels**. These channels can be expanded from the default ribbon to display each attribute of the item, such as the position, rotation, and scale. Of course, each of these channels can be edited in this mode (see Figure 4-8).

You now have the basic mechanics of how LightWave's interface is designed and how you operate the key features. I encourage you to open the LightWave proram on the free CD that comes with this book and explore these features in detail. Don't worry; you can't mess anything up permanently and you can always use the Undo and Reset functions to get back to where you might have erred.

Assuming you have spent some time exploring the LightWave interface, let's move on now to discreet's 3ds max.

3ds max

3ds max, or as most call it, just "max," offers a good balance between its depth of functionality and its difficulty to learn. It offers great character creation tools and a vast assortment of third-party plug-ins that seem to become incorporated as "included" features in subsequent upgrades.

General Description

Discreet's 3ds max presents all of its modeling and layout functions in one **user interface (UI)**. When starting, the animator boots to a default UI, but he or she may choose to customize the UI or select one of several configurations using the Layout tab.

3ds max's controls are iconic, which means that buttons are icons that represent their functions. If you park your mouse on an icon for a few seconds, the name of that button appears on the screen, making it easy to learn what the icons mean. Experts in 3ds max can customize their own toolbars and even evoke Expert mode (press Ctrl + X), which removes all toolbars and frees up more of the screen real estate.

3ds max can create a movie file format or an animation sequence of still frames that range from 1 pixel × 1 pixel to 9,245 × 9,245 pixels, which is far beyond what is currently required for cinematic projection. Such formats as D1 (NTSC and PAL), digital video, **high-definition TV (HDTV)**, and even large cinema formats, such as **70mm Imax**, are supported. Renders can also be output directly to compressed formats such as AVI and **Quicktime (MOV)** using any one of several **codecs** bundled in the software.

A clever new feature introduced in version 5.0 is the **splash screen** that appears while the program is booting. This screen differs for each boot, featuring one of more than 20 screens that feature a minitutorial on the program's available **hot keys**. I first thought the splash screen was just some clever graphic enhancement until I accidentally clicked one of the pictures and watched the tutorial play. What a good way to use otherwise wasted time to learn something.

A typical workflow in 3ds max would entail the following steps:

1. Create objects or import geometry
2. Add lights and cameras
3. Enhance objects with materials and effects
4. Build animations with keyframes

5. Enhance animations with dynamics (gravity, collisions, and soft bodies, such as cloth)

6. Add motion controllers and constraints

7. Enhance animations with modifiers (morphing, flex, and parametric modifiers)

8. Animate the cameras

9. Render and save image sequences or export geometry to other programs

Program Navigation

Unlike LightWave, Maya and 3ds max each combine their modeling and layout functions into one interface. You can, for instance, design a dagger, stick it into a table, decide that the dagger should be an ice pick, and change the blade from flat to round all in the same place.

The interface for 3ds max[2] is designed around **viewports,** into which objects are placed. When you start 3ds max, the default interface displays four viewports surrounded by tools and controls. A file with the suffix "cui" determines the colors, appearance, and layout of the UI. Several examples exist in the basic software and others can be created and saved for specific animators, projects, or functions.

At the top of the 3ds max interface is a row of constant hypertext buttons, known as the **Menu Bar**, which are similar to the standard category buttons of any Windows program. These buttons (such as File, Create, and Graph Editors) produce dropdown menus that represent the major categories of 3ds max operation.

Below the Menu Bar is a row of icon buttons known as the **Main Toolbar**. This is where the most common 3ds max tools (over 30, such as Undo, Select, Move, Render, and Materials Editor) are located. Some of the functions of these buttons replicate functions that are featured in the Menu Bar pulldowns but are here because they are used frequently (see Figure 4-9).

Another row of tab controls and a resulting group of icon buttons below the tabs are *hidden* by default, but they can be revealed between the Menu Bar and the Main Toolbar. This is the **Tab Panel** and can be revealed by

---

[2]Incidentally, I discovered several errors in the 3ds max tutorials that are delivered with the software, both on disk and in the printed book. Two examples: Right-mouse clicks are often written for what should be left clicks, and the view windows highlight in red, not in yellow, as indicated in the tutorials.

Figure 4-9

A common UI configuration in 3ds max showing the Tab Panel and Control Panel open, three orthogonal views set to wireframe, and one perspective view set to "Smooth and Highlights" shaded mode.

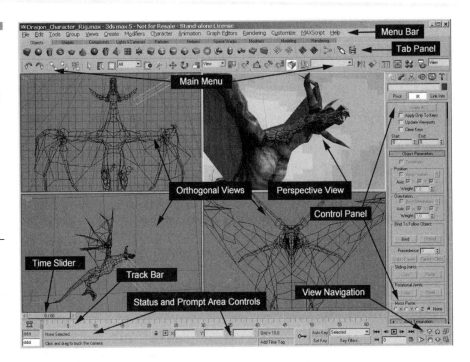

right-clicking any blank area of the Main Toolbar and selecting Tab Panel from the dropdown list. The Tab Panel controls the display of a selection of iconic category tools (such as Objects, Lights, Cameras, and Particles) just below the tabs.

Another menu tool reflects the current trend in animation programs to offer a text-based, schematic array of menu buttons produced under the mouse cursor on specific mouse clicks. Right-clicking any viewport will reveal these menus, which change based on what is selected in the view. Alternate menus are produced when the Alt key is held during the right-click (see Figure 4-10).

Below the viewports, at the bottom of the screen, are two horizontal rows representing the passage of time in 3ds max animations. The first row features the **time slider**, which can be dragged horizontally across the frames marked below in the **trackbar**. Dragging the time slider causes the read-out, representing the frame number over the amount of frames (such as 15/100), to change and for any animation represented in the viewports above to be activated.

Below the trackbar is another horizontal area of buttons and alphanumeric information windows. The most useful group in this area is the **status and prompt area controls**. The status bar displays what the operator

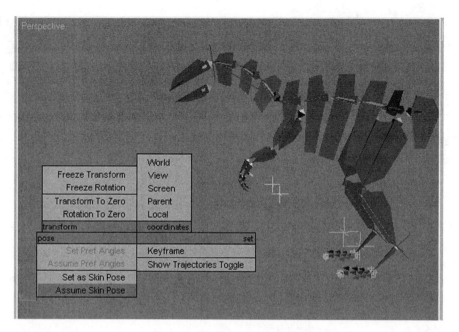

Figure 4-10
The right-click
menu revealed on
a viewport in 3ds
max set to the
perspective view

has selected, and the prompt line below that gives the operator a hint as to what the software expects you to do with that selection. The **Coordinate Display** windows, labeled x, y, and z, feature numeric readouts that correspond to mouse manipulations and enable numeric input from the keyboard (see Figure 4-11).

Also located in this area below the viewports are the buttons that control keyframe creation, the playback controls for the trackbar, and, in the far right of this area, the buttons that control view navigation.

To the right of the viewports is a moderately wide, vertical column called the **Command Panel**. The top of the Command Panel features five iconic category tabs (Create, Modify, Hierarchy, Motion, Display, and Utilities) that, when clicked, change the array of the tools and parameter controls below. The contents of each array are so extensive that the user can scroll and expand the display to two or more columns when accessing all the contents. Many of the tools found in the Command Panel are replicated in the Menu Bar.

With all the toolbars and panels opened at once, one might think that very little room is left for displaying the animation in the viewports. One might think correctly! That's why Expert mode has been created (toggled with Ctrl + X). Expert mode hides all the menus, buttons, and panels, revealing the real focus of 3ds max: the viewports.

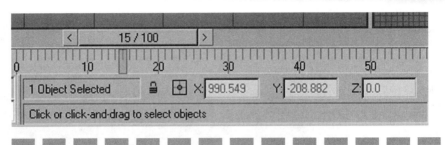

Figure 4-11
The time slider, a portion of the trackbar, and, below that, the status and prompt area showing that "1 Object" has been selected and that the program expects additional objects to be selected. Notice the coordinates that denote the current position of the cursor.

View Navigation

The four default views in *maxstart.cui* are top, left, front (orthogonal), and perspective. The view within any viewport can be assigned to the entire screen area by selecting the viewport and then clicking on the min/max toggle button in the lower-right corner of the screen. Viewports can display objects in wireframe mode, any one of several shaded modes, and an edge-faced mode that combines wireframe and shading. Each viewport can be assigned one of six orthogonal views (top, bottom, left, right, front, and back) or four perspective views (perspective, user, light source, or camera).

Right-clicking the viewport's label pulls down a menu where various viewing choices can be made. Here also, under Configure, is a utility that controls a wide range of viewport options, including the layout of the entire interface Clicking and dragging the mouse on the viewport borders adjusts the sizes of the viewports (see Figure 4-12).

The viewport views are manipulated in real time by using one of two **view navigation** button arrays in the lower-right corner of the screen (see Figure 4-13). These arrays change depending on the type of viewport selected. If the viewport is set to a perspective view, the controls reflect camera motions, such as dolly, field of view, and roll. If the selected viewport is set to an orthogonal view, the controls offered are limited to zooming (z-axis) and 2-D (x- or y-axis) moves. The two rightmost controls remain constant (zoom to full extent and full-screen toggle).

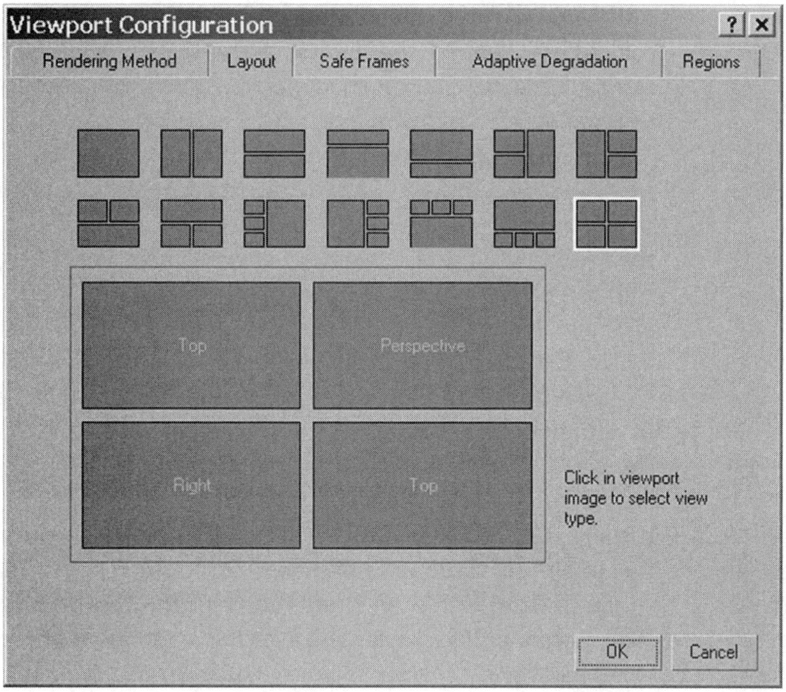

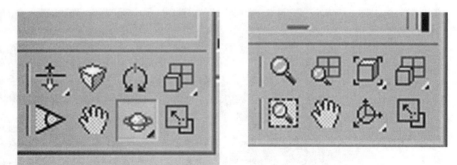

Figure 4-12
One portion (Layout tab) of the Configuration window that controls the viewing functions of the central viewports in 3ds max.

Figure 4-13
The view navigation constant button arrays in the lower-right corner of the 3ds max screen. Perspective view array is on the left, and orthogonal view array is on the right.

Object Navigation

Objects that can be manipulated in 3ds max include 3-D objects and elements, bones, lights, and cameras. Object navigation tools, found in the Main Menu toolbar as icons, include position, rotation, and scale.

When activated, object navigation tools superimpose the conventional colored symbols on the center of the object, relating to the type of tool selected. 3ds max, however, adds some interesting features to these symbols that are termed **gizmos**.

The **Move gizmo**, for instance, features a cube in the center of the arrows (see Figure 4-14). Grabbing the center of the cube allows movement in all axes, as you would expect, but grabbing one *face* of the cube, such as the back face, allows movement only on the two axes (x and y) that represent the plane of the back face. In other words, the movement of the object is constrained in the z-axis, but free in the x-y plane.

The **Rotate gizmo** includes a track ball embedded within the rings (see Figure 4-15). This ball is opaque to the rings, reducing screen clutter by blocking out the portions of the rings that go behind the ball. In addition, the ball's surface becomes the color of the chosen ring as you rotate the object. This highlighted wedge provides useful feedback, indicating how far the object has been rotated, as does a numeric pop-up window that displays the amount of rotation in degrees.

Similar to the feedback attributes of the Move and Rotate gizmos, the **Scale gizmo** also enables the user to choose planes of scale, and the gizmo

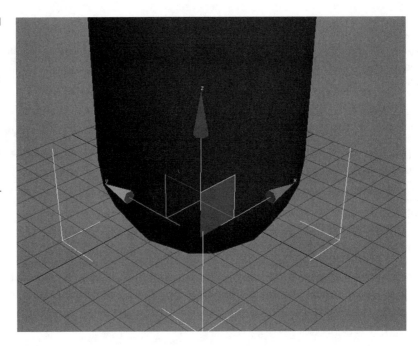

Figure 4-14
The 3ds max Move gizmo. Notice that the x-z plane in the center cube has been highlighted, indicating that movement on the y-axis is constrained.

Figure 4-15
The Rotate gizmo in 3dx max showing a rotation on the y-axis, displayed by a shaded arc around the rotation center point and the numeric readout above

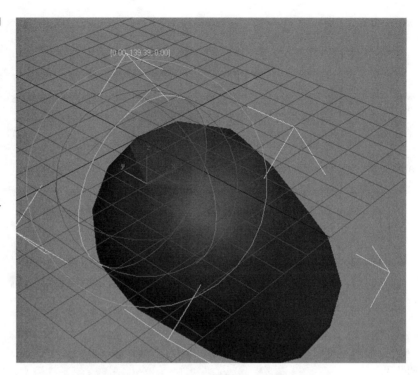

changes in size and proportion to the changes made on the object (see Figure 4-16).

A constant numeric readout in degrees is reported to the user within the x, y, and z windows of the **Coordinate Display** at the bottom center of the screen. At any time, you can click in these windows and supply precise numeric input from the keyboard.

Each gizmo can also be customized using a numeric window (Customize > Preferences > Gizmos) that lets you adjust the appearance, size, and functions of each gizmo (see Figure 4-17).

Object Creation

Points, curves, polygons, and solid objects are created in 3ds max by beginning with the Create button. From there, you can choose **standard primitives**, **extended primitives**, **shapes**, or **particles**. These choices and others under the Create button are also featured as separate tabs in the Tab Panel.

Figure 4-16
3ds max Scale gizmo shown with the x-z plane active. Note the bounding box, which is an optional display, is used to show the overall aspect of the manipulation.

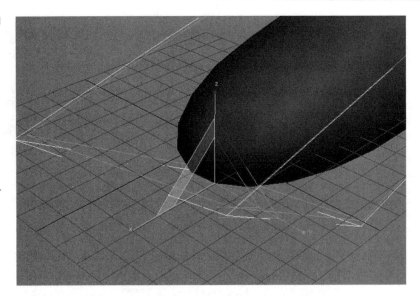

Figure 4-17
The Preference settings panel for the Move, Rotate, and Scale gizmos. Notice the wide range of settings designed to assist the operator.

Preference Settings

| General | Files | Viewports | Gamma | Rendering |
| Advanced Lighting | Animation | Inverse Kinematics | Gizmos | MAXScript |

Transform Gizmos

☑ On ☑ Show Axis Labels ☐ Allow multiple gizmos   Size : 40

Move Gizmo
  Relative Size (%) : 267.0

  Plane Handles
    ☑ On
      Size (%) : 100.0
      Offset (%) : 33.0

  Center Box Handle
    ☐ Move in Screen Space

Rotate Gizmo
  Relative Size (%) : 285.0
  ☑ Free Rotation ☑ Show Tripod
  ☑ Screen Handle ☑ Show Pie Slice
  ☑ Angle Data
  Rotation Method Circular Crank ▼
  Planar Angle Threshold : 10.0

Scale Gizmo
  Relative Size (%) : 207.0
  Uniform Handle Sixe (%) : 50.0
  2-Axis Handle Size (%) : 20.0
  ☐ Uniform 2-Axis Scaling

Move/Rotate Transforms
  ◉ Intersection   Persp Sens: 200.0   Rotation Increment: 0.5
  ○ Projection   Viewport Arc Rotate Snap Angle: 1.0

OK   Cancel

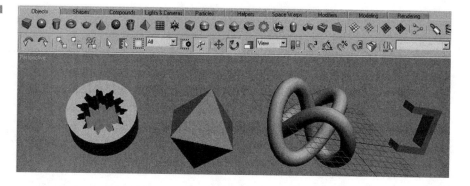

In addition to the wide range of primitives, including ring waves and torus knots, found in the Objects tab, 3ds max offers several other tabs with helpful tools. The Shapes tab provides the artist with basic 2-D modeling tools. These include point-to-point line-drawing tools and a generous assortment of line- and polygon-drawing tools, such as circle shapes, arc shapes, ngons (variable point-number polygons), text shapes, sections (as if sliced through a cone), helix shapes, and more (see Figure 4-18).

The Helpers tab offers a tool to make simple points, dummy objects, and some complex objects that are useful in architecture modeling, such as a compass rose, tape measure, protractor, and grid. Other tabs in this panel, such as Lights & Cameras, Particles, and Space Warps are advanced tool sets not covered in this chapter.

Object Editing

Object editing begins by clicking one of the object selection buttons in the Main menu. Five selection tools are offered (see Figure 4-19). The first is the Select Object pointer, which enables the user to click any object in the viewports. The second is the Select Object By Name button, which opens a window, listing all the objects in the viewports that the operator can select. The third tool is the Rectangular Selection Region, which enables the operator to drag a box in any viewport to select an object. The final tool is the Windows/Crossing toggle, which, when using the Rectangular Selection Region tool, enables the operator to decide if objects must be completely enclosed by the dragged rectangle or merely crossed by the rectangle.

Once the desired object has been selected, you can use a wide range of tools found in the Modifiers selection of the Menu Bar to execute changes. These tools are listed as groups, such as Selection Modifiers, Patch/Spine

Figure 4-19
The five selection tools in 3ds max, beginning with the Select Object pointer (highlighted), the Select By Name button, the Rectangular Selection Region, the Selection Filter pulldown menu, and the Windows/Crossing toggle.

Figure 4-20
The contents of the Modifiers list with the contents of the Mesh Editing submenu shown.

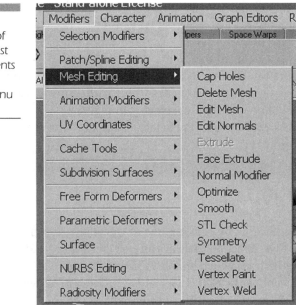

Editing, Mesh Editing, UV Coordinates, Subdivision Surfaces, Free Form Deformers, and so on (see Figure 4-20).

You can also activate the Modify panel from the Tab Panel bar to reveal a row of easy-to-use icons that provide a healthy assortment of pre-engineered effects. These include Taper, Bend, Stretch, Twist, and even some odd modifications like Melt and Ripple.

All modifier selections made in the pull-down menu or Menu Bar are represented as text and numerical statistics in the Modifier Stack Display with accompanying detail panels in the Control Panels to the right of the viewports. The Modifier Stack Display and the associated rollouts enable objects to be examined at a finer level of detail, known as the subobject level, which can be selected and edited with more precise control (see Figure 4-21).

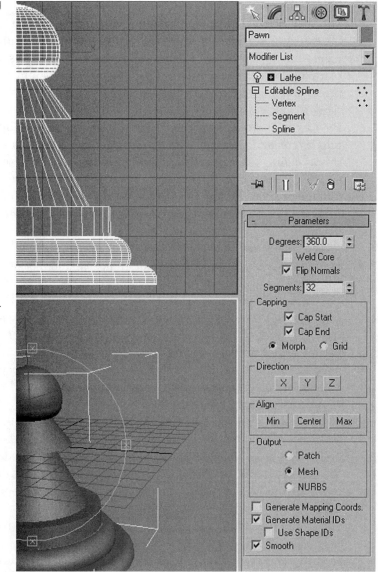

Figure 4-21

The Modifier List for the object, "Pawn," whose name is shown at the top of the Control Panel. Pawn's subobjects, Vertex, Segment, and Spline, are displayed just below. Under, "Parameters," notice the Lathe operation parameters that were used to create the Pawn from a curved line.

Layers

The function of layers in 3ds max is different from that of LightWave. In LightWave, the layers assist in modeling by enabling you to cut and paste parts of an object into different layers. In 3ds max, **layers** serve a different purpose. Here layers serve as a means of managing and grouping common objects within a complex scene, assigning certain common properties to those objects, and perhaps rendering those objects separately from objects in other layers.

Suppose you are modeling an office building. You might want all the marble exterior elements to be grouped in one layer, with all the marble elements having the same texture and reflective attributes. Then you might place all the glass elements, with the transparency and reflectivity appropriate for glass, grouped in another layer.

By Layer Properties By arranging various groups of common objects into layers, and slaving the properties of those objects to the layer (we call this assigning the properties *by layer*), you can automate the properties and speed up your work. For instance, if you wanted to see all the floors of the building without having to peer through the walls, you could hide the walls layer (do this by clicking the light bulb icon off) and all the walls would not disappear.

The properties of an object that can be controlled by layers are display, lock/unlock, rendering control, color, radiosity, and motion blur.

By Object Properties But what if you want to have Carrie-Anne Moss from *The Matrix*, dangling from a helicopter, smash into a set of windows while the rest of the windows in the building stay put? What do you do then? Certain windows can have their properties set up differently than the windows on the rest of the layer using individual object controls. Instead of being controlled *by layers*, the set of windows that are broken by Carrie-Anne crashing into them are set *by objects*.

You might also want to render all the computer-generated elements of this famous scene in one pass while you render the overlay of Ms. Moss in a different pass. This is a common workflow tactic, based on the fact that the actor's layer is already at maximum quality when it is digitally acquired and that the background elements may need several rendering passes to perfect. You wouldn't want to take time to render Carrie-Anne every time you tested the render of the glass wall, so you wouldn't render the actor's layer until the rest of the scene was perfected.

Every object created in or imported into 3ds max is assigned an initial layer. The default layer for all objects is layer 0, which has unique properties. Layer 0 enables random colors for new objects, and it is the layer where all imported objects are automatically assigned.

Finding the Layers function in 3ds max is a bit tricky. First, open the Layers toolbar, which is not visible by default, by right-clicking the Main toolbar (see Figure 4-22). Then choose Layers from the pulldown menu (see Figure 4-23).

Notice in the Layers Properties dialog box under the Name column, the first name listed is Layer 0. Under that layer are all the objects within that layer listed alphabetically. To the right of the object's name are clickable icons that represent the properties that can be "set by layer," as indicated by a dot, or that can be "set by object," as indicated by the pictorial icon.

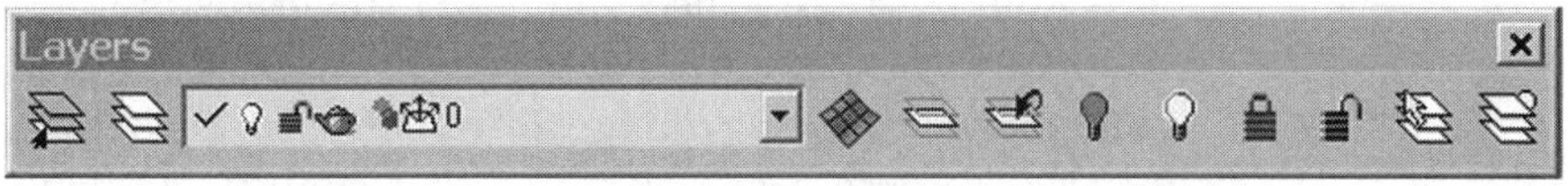

Figure 4-22
The Layers toolbar in 3ds max. The icon second from the left opens the Layers Properties dialog box.

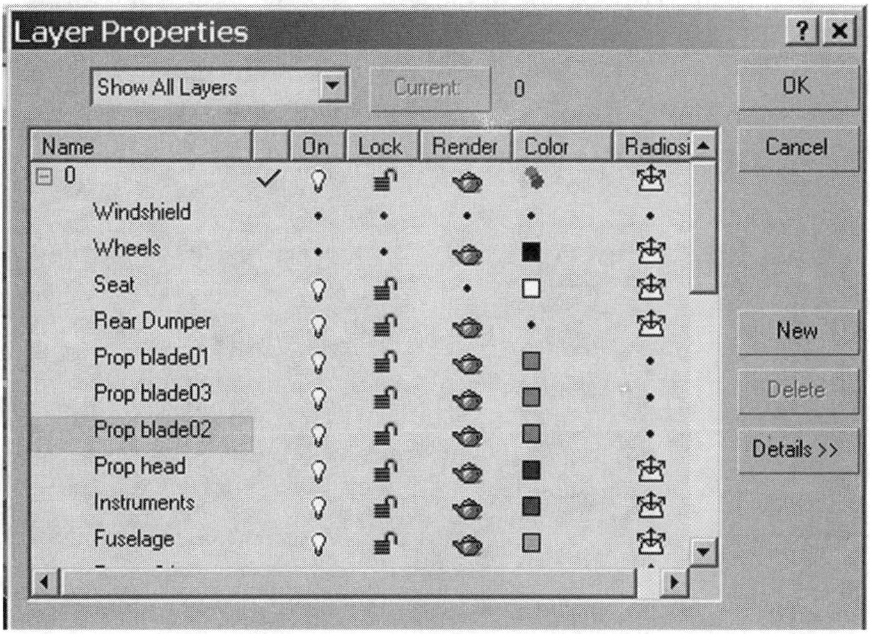

Figure 4-23
The Layers Properties dialog box in 3ds max.

Trackbar

The timeline in 3ds max is called the **Track Bar** and appears in a default length of 100 frames at the bottom of the interface window. Below the time-line are five VCR-type controls: play, frame step forward and backward, "go to beginning," and "go to end."

Animation is created by sliding the timeline cursor to a specific frame, manipulating an object or scene, and then creating a keyframe. 3ds max features an **Auto Key** button that creates a keyframe automatically in the current timeline position whenever a manipulation is made. Once keyframes have been added, the VCR keys can immediately be used to see previews of the animation in any or all views.

Here is a simple example of animation, adapted from one of the tutorials in 3ds max, where I have animated the orange to fall onto the table and bounce away (see Figure 4-24). Notice the keyframes that define the movement for the bouncing orange at frames 0, 10, 20, 30, and 40. Because frame 20 also features a one-frame scaling change (a slight y-axis squashing), frames 19 and 21 have been added to define a normal, spherical shape to the orange immediately before and after the impact. The orange is dis-

Figure 4-24

The 3ds max timeline is shown at the bottom of the screen with Auto Key on.

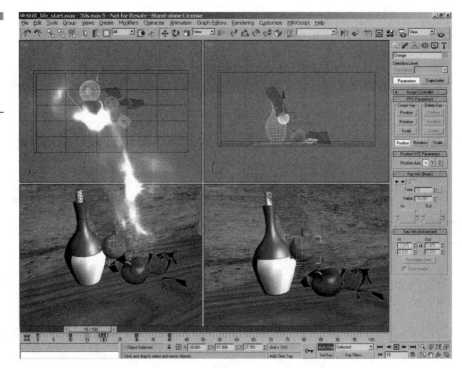

played in its current frame as well as several ghosted images of previous frames for the operator's reference.}

In this example, movement and shape are the two properties that have been animated, but you should know that any property of the object, its modifiers or materials, can be animated and in fact *are* in every scene. You may not see the animation, because every keyframe has precisely the same settings, so no change occurs. Therefore, you don't see anything getting animated, but the computer is paying attention to each and every attribute.

In 3ds max, you can see these attributes and their respective keyframes in two ways: **Curve Editor** and **Dope Sheet**.

Curve Editor   The Curve Editor features an expandable list of all a scene's elements and the attributes of each element (see Figure 4-25), such as the attributes of the orange I've named "Sphere01." The blue line in the graph to the right, for instance, indicates the change in the orange's Y-position from a starting position in frame 0 at +120 to an end position in frame 73 at -40. The graph elements, by the way, can be clicked and dragged to reset the attributes.

Dope Sheet   Okay, let's get all the jokes about dope out of the way, class, so we can get on with the lesson. The Dope Sheet is another way of looking at keyframes over time (see Figure 4-26). This method actually goes back to the oldest days of animation, when an animator would make a grid on a

Figure 4-25
The Curve Editor
in 3ds max

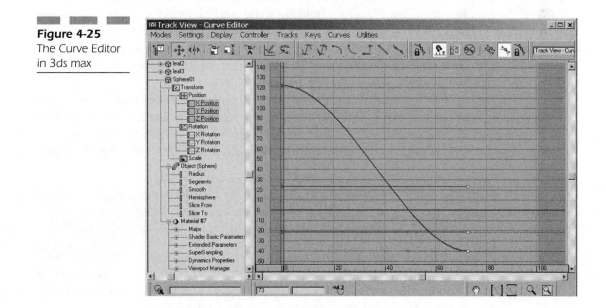

Figure 4-26

The Dope Sheet in
3ds max

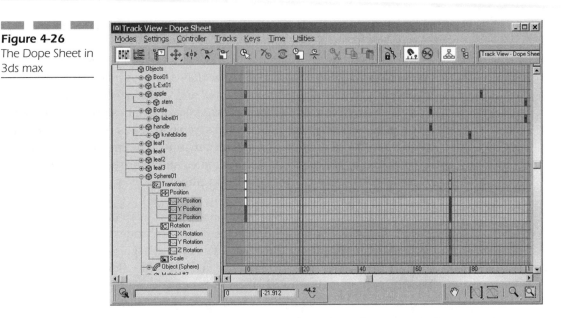

piece of paper with each horizontal line dedicated to an element in the scene and each vertical column being one click of the film camera's shutter (one frame). In order to keep track of things, the animator would start with frame one, click the shutter, and then fill in each block with a penciled mark.

For the next frame, the animator would move the objects in front of the camera, checking each one off the Dope Sheet until every object was accounted for. Then he would click the shutter to make the second frame. After years of experimentation, this was found to be the best way to make sure all the objects in a scene (and, for inked animations, every cell) were accounted for. Because any "dope" could make this system work, the grid became known as a Dope Sheet.

Now you too can use a Dope Sheet in 3ds max. Notice that the positions for the orange (X, Y, and Z) are expressed as filled-in boxes at Frames 0 and 73. The details of these boxes can be expanded by right-clicking each one, and as in the Curve Editor, the boxes are interactive with the Track Bar and viewports.

Exercise

Launch 3ds max, which opens in the default four-view workspace with Objects selected as the active tab. Then select the sphere object (second from the left). In any of the orthogonal views, make a sphere with a radius of about 40 (look under "Parameter" in the Control Panel).

Press the Set Key button at the bottom of the screen, and the Track Bar will turn red.

Click and drag the time slider from its default position at frame 0 on the Track Bar to frame 15. Click the Set Key button again, and notice that the tiny space for frame 15 has become darker. Now drag the time slider back to frame 0 and you'll notice that you've left behind a little orange marker on frame 15. This is your first keyframe.

In the Main menu, click the Move button (it looks like a cross with arrows on the ends). In the Front view, move the sphere above the x-axis (for now, try to keep it in sight in the Perspective view) and to the left of the y-axis. In other words, move it to the top-left corner of the view. Click the Set Key button again to make a new keyframe at frame 0.

Drag the time slider to frame 30 and notice how the sphere animates from the upper-left corner to the origin point. The sphere is beginning to animate. Using the move controls, place the sphere so that it is now in the upper-right corner of the view. Click the Set Key button again, and you should now have a keyframe at frames 0, 15, and 30.

Drag the time slider back to frame 0 and you will see your sphere animate backwards to the starting point. Press the Play button and you'll see your animation play. The default Track Bar length is 100 frames. Let's shorten it to 30. In the bottom right of the screen, look for the window that displays the current frame. Just to the right of this is a button with a square and a clock. Click it to open the **Time Configuration** window.

Where the End Time says 100, change it to 30 and click OK. The Track Bar length is now set to 30 frames. Press the Play button and the animation will continue to cycle until you stop it by pressing Pause (changed from the Play button).

Experiment with other forms of animation. If you feel adventurous, try elongating the sphere as it falls (to imitate an acceleration warp). Then squash it as it hits the surface, and elongate it again as it flies away. Now you're becoming an animator!

Take some time now to explore 3ds max using the trial version on the CD. Once you are satisfied that you have a reasonable grasp of the aspects described, move on to Maya.

Maya

Maya is considered by many professional animators to be the best animation program for choreographing characters. Curiously, Maya began as an imaging tool for the U.S. automotive industry, centered around Detroit.

General Description

Like 3ds max, Alias | Wavefront's Maya presents all its modeling and layout functions in one UI. The default UI features a large, perspective **scene window** with a **channel box**. Pressing the spacebar switches the large perspective scene window into a quad view of three orthographic windows (top, front, and side) and one perspective window.

Maya's controls are both iconic and textual. Iconic controls appear on a shelf above the scene window(s) and vertically along the left side. If you park your cursor over an icon for a few seconds, the name of that button appears onscreen, making it easy to learn what the icons do.

Hotbox A major navigation feature in Maya is the Hotbox, a text-based control interface. Whenever the spacebar is held down, the Hotbox, a hierarchical text menu, appears. While holding the spacebar, the mouse can be used to indicate a title in the Hotbox, which can open further menus. Menu choices in the Hotbox replicate iconic choices but often make the process of tool selection faster and more context-oriented than the pictorial rows of icons. The Maya operator can easily customize any of the icon controls or Hotbox contents, changing positions, shelf contents, or even the text or artwork of the icons.

Similar to the Hotbox is a contextual menu that appears in Maya whenever the right mouse is clicked and held over any object in a window. This produces a number of options presented as a hierarchical text tree that relates to the actions that can be performed on the object. This tool is a monumental time-saver.

Maya can create a movie file format or an animation sequence of still frames that ranges from 2×2 pixels to an excess of 10,000 pixels square, far beyond what is currently required for cinematic projection. Standard formats such as CCIR601, 720 and 1080 HD, PAL 760 and 780, 4k Square, and the usual **standard-definition** (SD) NTSC are supported. Renders can also be output directly to compressed **codecs** (**co**mpression **deco**mpression algorithms), such as **Cineon** and AVI, and sequential picture formats, such as eps, gif, jpg, iff, and tif are supported.

Hypergraph Hypergraph is a tool similar to the heirarchy display in LightWave that enables every element in the scene to be represented by a text box called a **node**. Each node can be linked to any other related node, easily displaying all relationships in a way that is familiar to anyone who has worked with an HTML authoring tool. Selecting a particular object or element is therefore facilitated. Groups of elements can be selected by holding Shift, and relationships between elements can be accomplished with simple drag-and-drop routines. The relative positions of nodes can also be moved on the display screen to represent their location in the scene. For instance, the nodes representing a lizard can be displayed as a head node, tail node, and nodes for the spinal segments and each leg.

For users of Maya that are familiar with Visual Basic, C, and C++, Maya offers a powerful tool, the **Maya Embedded Language** (**MEL**), which is capable of controlling any element within Maya. Although Maya's core programming is not changed by MEL, MEL can be used to automate functions, batch processes, create interactive menus, and even change the look and functions of buttons, menus, and icons. Advanced users of Maya tend to employ MEL in many ways, some of which become marketable plug-ins that advance the general capabilities of the program and can be offered to the community at large for profit or fame.

First-time users of Maya will be pleased to find that a collection of narrated movies appears after booting, including such basic topics as navigation essentials, creating and viewing objects, keyframe animation, and preview rendering. I think I'm in love with the woman whose voice narrates these clips. An extensive menu of HTML-based tutorials is also featured under the Help menu.

A typical workflow in Maya would entail the following:

1. Create scene objects or import geometry.
2. Add lights and cameras.
3. Enhance objects with Hypershade.
4. Build animations with keyframes.
5. Enhance animations with particles and soft-body dynamics.
6. Add motion controllers and constraints.
7. Enhance animations with fluid effects.
8. Animate the cameras.
9. Render and save image sequences.

Program Navigation

The interface for Maya is designed around a workspace of four basic **scene windows** or **panels** (top, bottom, front, and perspective) and two graphic windows (**Hypergraph** and **Hypershade**), as shown in Figure 4-27. When you start Maya, the default interface displays one large perspective window

Figure 4-27
Alias|Wavefront's
Maya workspace

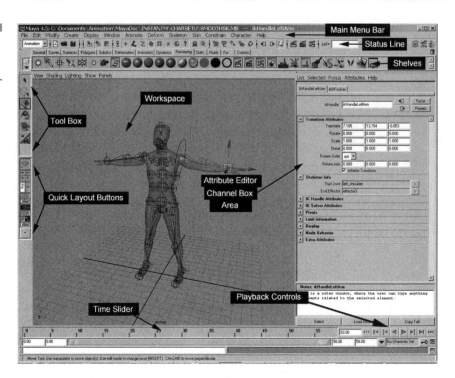

from the camera's **point of view** (**POV**), surrounded by tools and controls. This view can quickly be switched to a quad view by tapping the spacebar. Clicking the *middle mouse* (MM) button in any quadrant and then tapping the spacebar brings that quadrant to full size.

Layers

Additional view options can be selected from a group of seven **Quick Layout** buttons arranged in the lower half of the row of icons on the left of the scene window (see Figure 4-28). The first six of these buttons are tiny replications of how the interface will look when you press them. From top to bottom, these buttons are as follows:

- Default perspective view
- Classic quad view (see Figure 4-29)
- Vertical split screen between the outliner and a perspective view
- Horizontal split screen between a perspective view and the Curve Editor

Figure 4-28
The Quick Layout buttons as they appear on the left vertical border of the Maya interface

Figure 4-29
A quad orthogonal view in Maya. View parameters can be changed using the View Shading Lighting Show Panels menu at the top of each view.

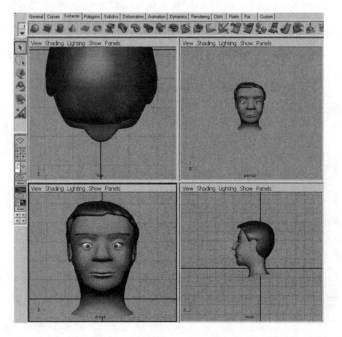

- Horizontal split between the Hypershade and a perspective view
- A three-screen perspective view shared with the Hypergraph and the Curve Editor

The seventh (bottom) button's appearance depends on which of the first six you choose. It divides itself into a group of buttons that represent each division of the chosen view. By clicking on any of these divisions, a menu appears in which you can change the content of that division to any other Maya display.

Just above the Quick Layout button array is another vertical array called the **Toolbox** (see Figure 4-30). The Toolbox contains the principal manipulator tools, such as Select, Lasso, Move, Rotate, Scale, Show Manipulator, and Last Selected.

At the top of the Maya interface are three rows of menus (see Figure 4-31). The top row, the **Main Menu Bar**, is a text menu divided into constant choices that do not change, including File, Edit, Modify, Create, Display, Window, and Help. The rest of the choices are changed by entering information in the window at the far left of the menu's row (or using the pull-down menu). The category choices are Animation, Dynamics, Modeling, and Rendering (**Maya Unlimited** adds another category, Live). When these are selected, the remaining choices in the Main Menu Bar change appropriately.

The next row of menus below the Main Menu Bar is the **Status Line**, composed of icon buttons that deal primarily with modeling and editing.

Figure 4-30
The Maya Toolbox can be found to the left of the scene window.

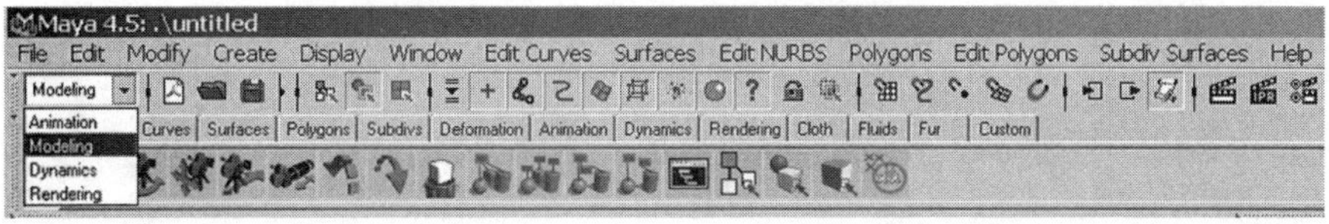

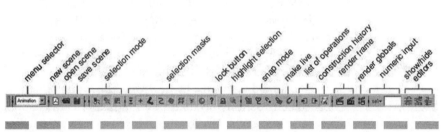

Figure 4-31
The three rows of menus at the top of the Maya interface. Notice the pull-down menu at the far left, indicating that the text choices of the top row of menus have been set to display modeling choices.

Figure 4-32
The groups of icons on the Status Line can be collapsed and expanded by clicking the narrow vertical bars shown between each function group. Graphic courtesy of Alias|Wavefront.

These icons are arranged in groups according to their function, as shown in Figure 4-32.

The bottom row of menus in Maya is called the **Shelf** (**or Shelves**), which the operator can customize by inserting and deleting iconic buttons (see Figure 4-33). The Shelves are controlled by a row of text tabs above the icons. By clicking each tab, a new assortment of icons is displayed. If the operator fills a given shelf with more icons than can fit, the shelf becomes scrollable, with the arrows at the far right of the display.

It is important to note that, in addition to the default icons in the shelves, the user can create additional icons with customized actions. For instance, if the user is animating an army of trolls, the creation of the troll's body can be assigned to an icon button and stored on a shelf. Thereafter, the "Troll Create" button can be pressed to execute a complex sequence of otherwise laborious and repetitive commands.

Keep in mind that all controls offered in the three rows of menus at the top of the screen can be replicated in the Hotbox and that both the Hotbox and shelves are independently customizable for maximum utility.

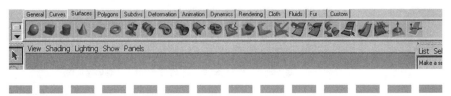

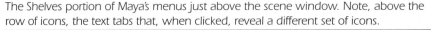

Figure 4-33
The Shelves portion of Maya's menus just above the scene window. Note, above the row of icons, the text tabs that, when clicked, reveal a different set of icons.

As you use the tools available in Maya to construct, modify, and manipulate objects and the scenes they compose, the software is busy in the background, keeping meticulous records of all the attributes you have assigned to every object. Many of these attributes are *keyable*, which means they can be assigned a value that changes over time.

You should be aware by now that the value of an animation program is based considerably on its capability to keyframe as many attributes of an object as possible. Obviously, in professional animation programs like the three in this book, the management of attributes that are keyframed can become quite complex.

In Maya, any element that can be keyframed is called a **channel**. For instance, if you have a car moving across the frame, the car can be said to be moving in the "x-axis channel" from point A (x = 0) to point B (x = 10). Of course, you may have hundreds or thousands of things happening in lots of channels during one short animation sequence. The complexity of channels in Maya is managed in a neatly organized text window called the **Channel Box**. The Channel Box is your central control area for all objects' keyable attributes. But what about the nonkeyable attributes?

Another larger box called the **Attribute Editor** holds both the keyable and nonkeyable attributes. Because the Channel Box is a subset of the Attribute Editor, the two boxes swap space to the right of the Main Scene window by pressing Ctrl + A (see Figure 4-34).

Although the Attribute Editor may look quite daunting due to its incredible array of features and terminology, you will, with practice and tutorial work, soon become familiar with all its complexity. For now, you may want to experiment with the simpler Channel Box tools.

Exercise

Load an object into the Main Scene window or create an object by going to Create > Nurbs Primitives. Click on the object or, if you've loaded a complex

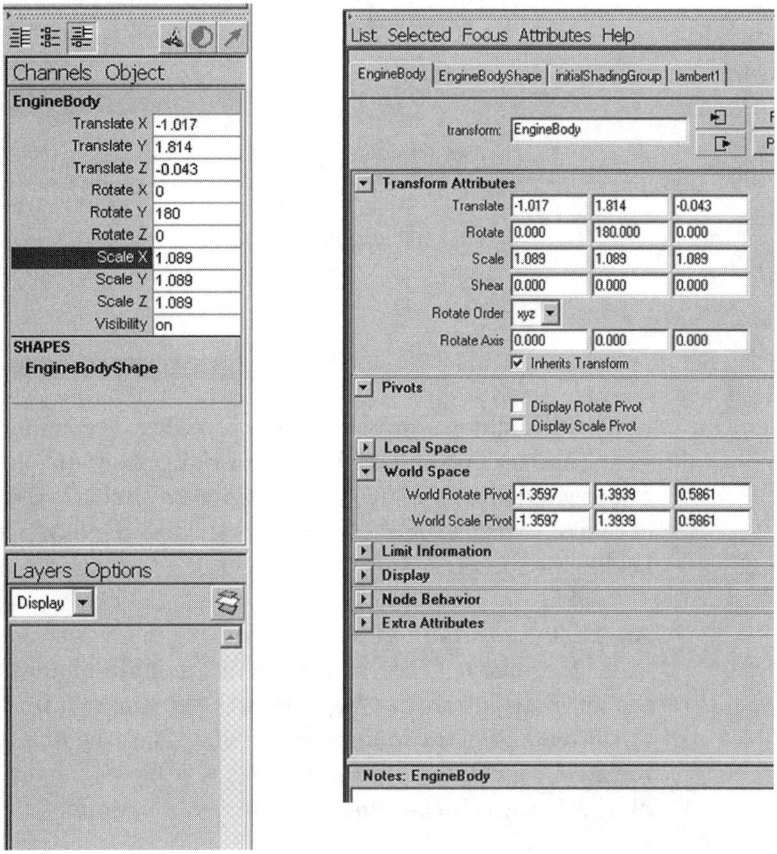

Figure 4-34

The two boxes that contain all the attributes of any object in Maya. The Channel Box, left, and the Attribute Editor, right, are toggled by Ctrl + A.

object, click on a portion of it to highlight it. In the Channel Box, select any attribute, such as Scale X, by clicking on the text Scale X.

You'll notice that Scale X is highlighted in black and that within the object a three-axis scale tool appears. Obviously, you can now move your cursor into the Main Scene window and drag on any scale handle to affect the scaling of any axis. But if you click and hold your middle mouse button in the Main Scene window, the three-axis tool becomes a simpler one-axis slider that constrains the effect of your cursor to the x-axis. When working with complex structures, this feature of Maya can save you a lot of time.

Below the main scene window, extending across the width of the workspace, are the time slider and playback controls (see Figure 4-35). Like LightWave and 3ds max, Maya uses the time slider to control animation and store keyframes. Two windows on either side of the slider enable the user to numerically set both the beginning and end frames of the animation, as well as the start and stop points of a play sequence within the animation. To the

Figure 4-35
The time slider
and playback
controls in Maya
extend across the
bottom of the
workspace.
Illustration
courtesy of
Alias|Wavefront.

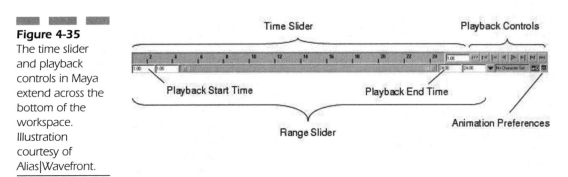

right of the slider is an assortment of VCR controls for playing the animation. The last button on the bottom right opens the **Animation Preferences** window. The controls in this window enable you to open a wide range of controls within four categories: Interface, Display, Settings, and Module loading preferences.

Another Exercise   First, create a cube in the workspace. Drag the time slider to frame 0. Click Animation in the Main Menu tabs and then press the leftmost button, Set Key. Move the cursor to frame 30. In the Toolbox on the left side of the Main Scene window, click the Rotate button and rotate the cube significantly. Press the Set Key button again. Press Play on the playback control. The cube will rotate as the animation loops.

View Navigation

The easiest way to manipulate views in Maya is to simply press and hold the Alt key and use your mouse button. Pressing Alt with the middle mouse button will pan and tilt any view. Using Alt and the right mouse button will zoom the view in and out.

In the perspective views, you can tumble using Alt with the left mouse button to rotate the scene (moving the mouse left and right will move the view up and down on the y-axis, while moving the mouse forward and back will move the view left and right on the x-axis). Of course, orthogonal views cannot be tumbled, because by definition they represent static perspectives.

Any view can be quickly toggled to a close-up view by pressing the *f* key. Pressing the spacebar toggles the Main Scene window from the default one-view perspective view to a four-view quad (top, front, side, and perspective).

Object Navigation

Similar to LightWave and 3ds max, Maya enables the keyable manipulation of all 2-D and 3-D objects, as well as elements, bones, lights, and cameras. The three **manipulators** are move, rotate, and scale, and the buttons to select each of these are within the Toolbox to the left of the main scene window.

Activating each manipulator reveals an appropriate symbol within the object, similar to those found in LightWave and 3ds max. The symbols have color-coded end handles that correspond to the axis of manipulation controlled by the handle. For all manipulators, the x-axis is red, the y-axis is green, and the z-axis is blue. Grab the handle to select the manipulation of your choice or grab the central point of the symbol to freely manipulate the object.

After selecting groups of objects, simultaneous manipulations can be made on the entire group by using the symbol on only one member of the group. Conversely, by selecting only a portion of an object (termed an **element** in Maya), just a portion of the object can be manipulated, such as bending, twisting, or stretching a portion of the object to create a new shape. Numeric readouts of the manipulations in Maya can be found in the Channel Box or Attribute Editor to the right of the main screen window.

Object Creation

Under the Create menu, Maya provides four categories of primitive shapes: **NURBS** (or **Nonuniform Rational B-spline**), **Polygons**, **Subdivision Surface**, and Volume (see Figure 4-36). Within each category is an assortment of standard shapes. Clicking the name of the shape produces a default-sized object of that type at the zero point of the x-, y-, and z-axes, known in Maya as the **origin**. In order to customize the object before it is created, click the box to the right of the name of the object. This opens up an option box with parameters that can be defined numerically and with sliders (see Figures 4-37 and 4-38).

In addition to the pull-down menus under Create, the **Polygon shelf** offers a default assortment of primitives, whereas the **Subdivs tab** produces a limited range of polygons with subdivision surfaces. Other tabbed shelves offer iconic buttons that enable the user to create curves and surfaces, and to deform surfaces and objects (using the **Deformation shelf**) in

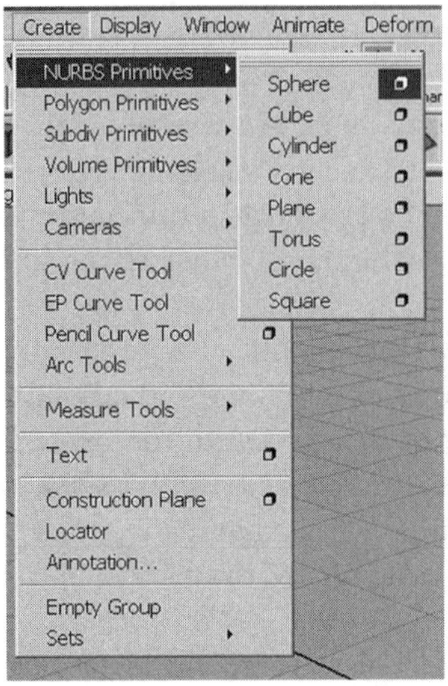

Figure 4-36
The Create menu, with the NURBS category selected. Clicking the box to the right of Sphere will open an attribute box.

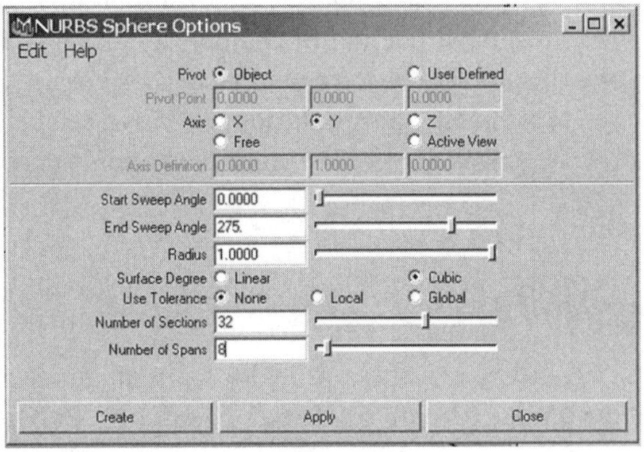

Create	Display	Window	Animate	Deform

NURBS Primitives ▸

Sphere	▢
Cube	▢
Cylinder	▢
Cone	▢
Plane	▢
Torus	▢
Circle	▢
Square	▢

Polygon Primitives ▸
Subdiv Primitives ▸
Volume Primitives ▸
Lights ▸
Cameras ▸

CV Curve Tool
EP Curve Tool
Pencil Curve Tool ▢
Arc Tools ▸

Measure Tools ▸

Text ▢

Construction Plane ▢
Locator
Annotation...

Empty Group
Sets ▸

Figure 4-37
The attribute box with settings changed to express a sphere with a 275-degree sweep angle in 32 sections and 8 spans

NURBS Sphere Options

Edit   Help

Pivot ⦿ Object ◯ User Defined
Pivot Point 0.0000 0.0000 0.0000
Axis ◯ X ⦿ Y ◯ Z
◯ Free ◯ Active View
Axis Definition 0.0000 1.0000 0.0000

Start Sweep Angle 0.0000
End Sweep Angle 275.
Radius 1.0000
Surface Degree ◯ Linear ⦿ Cubic
Use Tolerance ⦿ None ◯ Local ◯ Global
Number of Sections 32
Number of Spans 8

Create	Apply	Close

Figure 4-38
The actual sphere created by the inputs in the option box

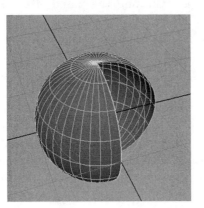

various ways. As noted before, the shelves contain an adequate array of default choices, but the user may easily add, remove, or change these choices.

In Maya, the number keys can be used to quickly change the display, but not the actual shape of an object. The 1, 2, and 3 keys display ever smoother edges of NURBS objects, whereas the 4 and 5 keys toggle between wireframe and shaded mode.

Object Editing

Until this point, we have been working in **Object mode**, where selection tools and modifiers have been applying themselves to an entire object or groups of objects. In order to edit an object, that is, to take control of an object's parts or **components**, you have to switch into **Component mode**.

Exercise

Create or load a primitive polygon object, such as a sphere, into the main screen window. Place the cursor over the object and click the object to select it. Click and hold the right mouse button to reveal the **Marking menu**. Drag the cursor to the selection face. Release the mouse button and you'll see the object displayed in Component mode. Click any face, or click and drag to select more than one face. In the Tool menu, select Move. Drag any handle on the Move gizmo to displace the face(s) you selected. Use Ctrl + Z to undo any actions you don't like and try again. Try different object types and experiment with various manipulators until you get comfortable with Component mode.

The selection of components in Maya follows the Windows convention of using the Ctrl and Shift keys. When selecting elements, holding Shift while clicking additional elements will enable you to add elements to those already selected. Holding the Ctrl key and clicking elements that have been selected will deselect them.

Layers

To access the layers' control, use the Main menu selection: Display > UI Elements > Channel Box/Layer Editor. Layers are used for two separate purposes in Maya: display and rendering.

The **Display Layers** function enables the user to place objects and elements into different, color-coded layers that can be made visible or invisible in different views. Breaking up a complex object or scene into various layers assists the animator in identifying and modifying objects and elements. Care must be taken when building the complex object or scene to allocate the layers and their contents so that the resulting organization makes sense and is easy to navigate. Once organized, a complex array of layers can save hours of hunting, grabbing, and releasing unwanted targets. An example might be a starship used in a television series. The master ship model may not be needed for every scene, and if the animators need to isolate the shuttle bay interior to build a new scene in that space, they can access a specific layer that contains only the shuttle bay.

The **Render Layers** function enables the user to render the contents of a defined layer or layers without rendering the contents of others. This is useful in test rendering a complex scene, where the operator wants to isolate a particular aspect for rendering without wasting time rendering all the scene's contents. An example might be a smoke effect running through a complex city scene, where the animator wants to see how the smoke particles render, without wasting time rendering the buildings, cars, streets, or people.

Exercise

You can create new layers by clicking Layers > Create Layer at the top of the Layers Editor box, or simply click the Layer icon. You can name the new layer by double-clicking it, which opens a dialog box that enables you to rename and color the layer. Now you can add objects to this layer.

Now create a few objects in the main scene window and then select one. Select Layers > Add Selected Objects to Current Layer in the pull-down menu that appears, or click and hold the layer name to produce the same menu. The object you selected is now stored in the new layer. Return to the main scene window and deselect the object. Now, to demonstrate how you can easily reselect the object without touching the main scene, go to the layer name, right-click and hold on it, and choose Select Objects. Presto! The object is reselected.

If you now change Display to Render in the Layers definition window, you'll see that no layers are named or selected. This is because the Render Layers function is completely separate from the Display Layers function and holds a totally different database of contents.

Back in the Display Layers list, you'll notice three buttons to the left of the layer name. The button with the "V" is a toggle that makes the layer visible (V) or invisible (blank). The middle button toggles one of three display modes: reference (R), template (T), or normal (blank). Reference shows the outline of the object but won't enable the object to be selected or modified, but you can *snap* to it.[1] Template acts exactly like Reference, except it prohibits the snap-to function. Template and Reference are useful for keeping an object, such as a bridge, visible in the work screen while you design something that fits the template, such as a car. The colored button displays the color you selected for the layer. When double-clicked, it opens the Edit Layers dialog box (see Figure 4-39).

The Render Layers feature is much simpler than Display Layers. Although the methods of creating, naming, and assigning objects to layers are identical to the Display feature, only one function button exists with one toggle. An R (Render) in the button means that object in the layer will render. A blank button means it will not.

Time Slider

Like most 3-D animation programs, the timeline in Maya (called the *time slider*) resides at the bottom of the screen under the Main Scene window.

Figure 4-39
The layers control area in Maya. Note Display in the choice window and the three buttons for each layer. Clicking the layer name opens the Edit Layers box shown on the right. Images compliments of Alias|Wavefront.

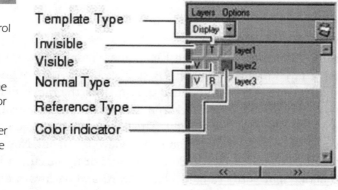

Template Type
Invisible
Visible
Normal Type
Reference Type
Color indicator

---

[1]Snapping is a common feature that creates a kind of magnetic effect between two objects in a UI, thereby making it easier for an operator to place a movable object next to an anticipated target. In this case, other objects can be snapped to referenced objects.

The Maya timeline is composed of two horizontal bars and five numeric windows. The top bar is called the time slider, which displays specific frames within a set range (the default range is 24 frames). At the far right of the time slider is the Current Frame numeric window.

This window displays the frame that is being shown in the main scene window. The current frame is also indicated in black, which in a zoomed-back view will appear as a black vertical line, but in an extreme close-up, it appears as a black range equal to or less than the width of one frame. This is the time slider cursor, which can be dragged across the time slider to cycle through the animation to and from specific frames. The user also can also click in the Current Frame window and type a specific frame number. Note that Maya supports decimal fractions of frames as well.

The bottom horizontal bar is the range slider. Within the range slider is a movable cursor. The horizontal length of the cursor represents the relative width of the time slider window. By dragging on the handle at either end of the cursor, the width of both the cursor and the corresponding view of the time slider can be adjusted. Once adjusted, the cursor can be dragged left and right to target the time slider within a specific range of the entire animation. This is a useful tool when working on animations that are longer than the window view.

On each end of the range slider are two numeric windows. The windows nearest the slider indicate the playback beginning and end frames. The outside windows define the beginning and end times of the entire animation.

Exercise

Launch Maya and click in the End Time window (farthest to the right). Enter a large number, such as 1800. Notice how the range slider cursor shrinks (because you've expanded the timeline, the range slider now represents a smaller part of it). Click and hold on the right control handle of the cursor and drag it left and right. This action changes the gradations and numbers in the time slider, zooming the view in and out. Now click and hold in the middle of the range slider cursor and drag the cursor left and right. This action moves the range slider forward and backward along the timeline. Now click once in the time slider window on any frame number or gradation line. This places the timeline cursor on that specific frame. The specific frame number you selected will be shown beside the cursor on the lower edge of the time slider.

Additional controls and presets for the timeline and playback parameters can be found in a popup menu by choosing Window > Settings/Preferences and then clicking the heading title Settings and the subtitle Timeline. This window enables you to enter some of the same numeric values as the UI timeline as well as additional entries for the timeline track height (1x, 2x, 4x), which is useful for seeing and synchronizing to imported sound files. Here you can also set the **playback speed** (every frame, every other frame, every minute) and the **playback by** rate (Maya's default is 24 *frames per second* [fps]).

Playing back animations is simple. Just use the VCR controls at the right of the timeline or use Alt + V to start and Alt + Shift + V to instantly rewind the video to the first frame.

Summary

This chapter provided a basic explanation of how three leading animation programs work. I encourage you to explore each one using similar tutorials. Which program works best for you? This is the time to decide.

Later, as you commit yourself to more learning time and practice, you will settle in to one program as your favorite (no one stays a master in all three). When that time comes, it will be difficult to change your mind and begin to master another program. But you can remain familiar with the other programs and use them occasionally as tools to achieve a specific goal, which may be more difficult in your favorite program.

As you learn, the programs themselves also develop and grow, changing from what you originally experienced. Frequent exploration, using free downloads, reading magazines and web info, and attending trade shows like **SIGGRAPH**, will allow you to keep up with other programs to see the upgrades they've made.

As you explore the software and learn to animate, don't forget to keep making contacts and sales calls. Often the demands of your customers will tell you which animation program is in demand and what you should be learning.

The next chapter continues the exploration of the animation craft by exploring objects. When you've exhausted yourself on the software, get back to reading!

Objects and Surfaces

This chapter explains how humans and computers interact to create a 3-D object and its surfaces. You will learn how the surfaces of these objects can be assigned different attributes that give them color, texture, and shape.

The process of computer animation can be broken down into two halves: creating objects and animating objects. Some programs, such as LightWave, divide these functions into separate programs (Layout and Modeler). Others put everything together.

When creating objects, we must consider two key aspects: the object's shape and the surface's attributes. Whichever program you are using, the best place to begin your actual exploration is by building objects and understanding their various attributes.

Object Shapes

All objects are created in a virtual 3-D space, which is defined by a 3-D grid known as a **Cartesian plane**. Although the size of the plane can represent any dimension from the microcosmic to the cosmic, it is best, before you begin, to decide on the scale your work will have. For instance, if you are going to build a car, you would choose a grid size based on a maximum view of 10 meters.[1] If you were going to design a star cruiser, you might want to work on a larger scale, but only so large as to accommodate all components. Computers tend to initiate rounding errors as large relative scales.

Each program has a means of specifying the default grid square size. Frequently, animators design objects in a 1:1 scale. That means if a boat is 50 meters long in real life, it is designed 50 meters long in the animation program. This is useful when matching your objects up with other people's work or if you want to sell your objects to a stock object house or on the Internet. Keeping a 1:1 scale keeps everyone in synch.

An important concept to keep in mind while creating objects is how they will be used in the end medium. Many people study three-dimensional animation programs, particularly the object-modeling routines to build prototypes for real objects, such as a new car, an interior residential space, or a building. Objects designed for eventual application in the real world must leave the computer realm in a more complete form than that required for animation. When creating for movies, video, web sites, and other 2-D work, however, an animator creates only the portions of the objects that will be

---

[1]With some exceptions, I will use the metric system throughout this book. Because metrics are always divisible by 10, they are the easiest to scale up and down.

seen. For instance, if you plan to design a car that will drive past the viewer from left to right and never be seen in any other fashion, there would be no need to create the driver's side. The same goes for characters that never turn their back on the audience. Obviously, it helps to carefully plan the requirements of your animation before you start the modeling process. This process begins, of course, at the storyboard phase, where you determine exactly what you expect to show.

By the way, an excellent 3-D storyboard program you might wish to try is Frameforge 3D Studio from Innovative Software, LLC (www.FrameForge Studio.com).

Most frequently, animators choose one of the three orthogonal views in which to begin building an object. Because most objects are intended for a frontal view, the front view is most frequently chosen, but often simple objects are combined with more complex objects. In such cases, the added objects can be created and merged with the original object using the top, left, right, or other views as necessary. The perspective view, which can be kept open on the workspace, is used to survey the results, and animators will often move the view around after each change to the object.

 ## Why Model Objects at All?

After more than a decade of 3-D animation, you might guess that someone somewhere has already created nearly every object imaginable. You're pretty much right. One day you might be asked to model a 1959 Chevrolet Impala. You could certainly buy a plastic model at a hobby store and start to work, or you could buy the object, already made, from a data warehouse!

As mentioned earlier in Chapter 2, "From Concept to Screen: The Work-flow Process," **Digimation** (www.digimation.com) buys and sells 3-D objects, edited for every available software program and computer capacity. Data warehouses can be found on the Internet or in trade journals. Their catalogs, filled with the most amazing collection of objects, from human body parts to World War II battle tanks, are fascinating viewing. Another source of 3-D objects is **Turbosquid** (www.turbosquid.com), which acts as an agent for animators interested in buying or selling objects.

Expect to pay a significant fee for objects when purchased from a data warehouse. A human figure, for instance, designed for 3ds max, with 66,800 polygons, will cost a bit over $1,000. The good news about buying objects from a reputable source is that the objects are clean, ready to use, format-ted for your software, fully tested, and free of viruses.

Keep in mind that you can also sell your objects to data warehouses. Once you've finished designing a specific object, consider its generic applicability in the marketplace. Could someone else conceivably use the same object? If you are the author of the object, you can sell it outright or on a commission basis to a data warehouse and perhaps over time make some handsome profits.

Internet Objects

Exploring the Internet will also yield several free or **shareware** sources of objects offered by generous enterprising animators. **Free objects** are just that. Take what you want, use them how you will, and don't worry about paying anyone. Shareware implies that you owe the original designer a fee, usually quoted somewhere on the offering page. Shareware is an honor system. You can download the object for free and use it how you will; you are then are honor bound to send the designer the fee quoted.

You may find just the object you need or something close enough that will work with a little bit of tweaking (the computer kind). If you're lucky, you can save yourself days of work. Often, you may find that the object you want is in a format made for a different animation program. No problem. Your program may offer an object format conversion module, or you can purchase one from third-party vendors.

All object formats are based on simple files composed of text characters. Because all text characters come from the universal **American Standard Code for Information Interchange** (**ASCII**) character set (the same set used to assign specific characters to each of your keyboard's buttons), converting an object from one format to another is rather simple. You may not retain all the intricacies of design that the original software program imparted on the object, but you will most probably retain the original points and wireframe.

Modeling Methods

In the earlier days of animation, the only way of creating complex objects was by manipulating and combining groups of flat polygons. This method came to be known as **polygon modeling**. The drawback of this method was that it tended to create objects with prominent angles that looked boxy and artificial. To avoid the boxiness, the component polygons were subdivided into smaller elements until the edges tended to smooth out and disappear.

Later, more advanced methods, based on modeling with curves, were invented, such as **splines**. Although mathematically more complicated than polygon modeling, splines allowed for attractive, rounded surfaces, composed of nonflat polygons so that they could be modeled easily. Splines and other methods that were invented later attracted a certain number of animators, creating groups that would prefer one method over another. Even later, refinements to polygon modeling and spline modeling were introduced, reviving the interest in earlier techniques and further complicating the factions.

Polygon Modeling

We'll begin by describing **polygon modeling** and proceed to other methods. Each of the three animation programs in this book have different ways of offering users the various methods of modeling. Some programs, such as LightWave, handle the various modeling methods differently. Although specific differences will not be covered in detail in this book, you are encouraged to explore them in the tutorials provided by each manufacturer on the free CD.

Simple objects, such as spheres and cones, are made more complex when the operator selects groups of points or polygons and moves or extrudes them. For instance, a single point on a sphere can be selected and pulled away from the center to form a pointed extension of the sphere, such as the tip of a nose (see Figure 5-1). Selecting a wider range of polygons, pulling them away from the center, and then rotating them slightly downward might make a more elegant nose. In LightWave, tools like **Drag**, **Magnet**, and **Dragnet** provide this capability. In 3ds max, you might use **Editable Poly**. In Maya, you use **Manipulator Handles**.

Animation programs offer a means by which points and polygons can be selected and manipulated to create more complex shapes. For the novice, this is always an activity that goes well for a few steps and then falls apart. I would suggest, while experimenting with this technique, to start with a basic object and save it with a simple name, such as "head." Then make your first adjustment. If it looks bad, click undo and start over. If it looks good, save the result immediately as head1 and proceed to the next adjustment. If that object looks good, save it as head2 and continue.

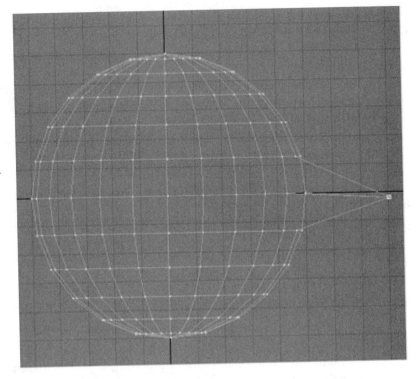

Figure 5-1
Example of a
sphere with one
point selected and
dragged out from
the center.
Perhaps the
beginning of a
cartoon head's
nose?

Some animation programs now feature a **history of actions** list that can be accessed to backtrack every action you executed. The most experienced animators, however, tend to fill up their hard drives with similar renditions of objects, some representing various design tangents that they took on the way to perfection. If each critical decision is saved, the animator merely refers back to his or her list of named objects and backtracks. I have objects named "FaceThatClientLikedonJan5ButIhated" and "CarAfterSmashOnLeftsideBeforeAdditionOfHeadlights." Thank heaven Windows acquired the capability to name files with more than eight letters!

Points and Polygons At the most basic level, defining points and polygons creates an object. The process is very simple: Define at least three points somewhere in the scene and connect them with lines to form polygons. In animation, the most common polygon is the triangle, because it will always serve as the basic element of the most complex surface. When divided small enough, large arrays of triangles can be used to model the most intricate details. In addition, the triangles also have a unique property. Imagine a triangular piece of plastic laying on a table. If you lift one corner, the other corners stay flat on the table. You cannot do this with any other polygon. This unique attribute allows triangles to compose any surface and allows that surface to flex in any way imaginable (if the triangles are small enough).

Animation programs have various tools for changing squares and irregular polygons into subdivided arrays of triangles, and you will no doubt use them frequently. These arrays are often called **meshes**. Keep in mind, however, that the more polygons an object has, the more memory it will take to describe and the longer it will take to render. Again, it is important to anticipate the end use of your object and design it appropriately. If you are making a simple cube that will inflate with air and become rounded, you will need to subdivide the surfaces with enough triangles (perhaps hundreds) to create the illusion of roundness. If the cube will remain a cube, however, you can get away with six points and six planes, or even fewer if the cube does not rotate.

Obviously, a polygon can have any number of points. When designing a logo, for instance, you might find that a complex shape, such as the letter P, can have dozens of points (sometimes referred to as **n-gons**). If all the points lay on one plane, you will have no problem defining the surface within those points, and the computer will render that surface predictably. If, however, you should cause a polygon with more than three points to bend, you could easily move one or more points off the common plane. This would define a surface, which is difficult, if not impossible for the computer to render.

Although you may think otherwise, computer animation allows the existence of one- and two-point polygons. A single-point polygon with appropriate attributes can serve a number of useful purposes. It can be invisible (called a **null object**) and serve as a reference or **parent** object to which other objects are subordinated. It can also be visible and serve as an infinitesimal source of light, such as a star.

Two-point polygons define a line, such as a string or hair strand. Under most conditions, two-point polygons always remain thin, however, no matter how close the camera zooms into them. Many programs use two-point polygons to define complex arrays of strands, such as a field of grass, a head of hair, or rain.

Primitives   Animation programs offer a default assortment of primitive 2-D polygons and 3-D objects that can be created with a few clicks of the mouse. Polygons such as triangles, squares, and regular polyhedrons (pentagons, hexagons, and so on) can be quickly brought to the scene by selecting the appropriate tool and defining the boundary of the polygon in one of the orthogonal views. In addition to the click-and-drag option, each program offers the capability to enter numerical values and, after clicking the Create or Apply button, the computer creates the polygon precisely as defined.

Choosing from an assortment of buttons can easily create primitive 3-D objects such as cubes, spheres, pyramids, cones, and so on (see Figure 5-2). Once the type of object is selected, the operator chooses a view, places the cursor, and drags out a boundary. Letting go of the cursor initiates creation of the object.

Numerical Entry   Numerical entry is a process of specifying values by typing numbers on your keypad, as opposed to dragging object components on the UI. Numerical entry is available at all levels of animation production and is useful in maintaining consistency that cannot be achieved with click-and-drag input. Although your program will create any primitive

Figure 5-2

Example of several primitives quickly created in an animation program

using a default selection of parameters (for instance, a sphere can be created using a preselected amount of radii), you can change these defaults using the Preferences tool or set the parameters for each object prior to creation. Novices are encouraged to try creating a few primitives and examine the results. Then change the default preferences and try the same primitives again. Eventually, you will get a sense of which parameters are best for your needs at any given time. For example, if you are building very elemental shapes, you can set the parameters at the minimum. If you are pulling out primitives to build a complex object, however, such as a sphere for a human head and a cone for the nose, you will select a higher number of subdivisions, knowing that eventually you will need the complexity for shaping the details.

Numerical input can also be useful in other areas of object manipulation. For example, you may have designed an object and want to rotate it exactly 180 degrees (half of a symmetrical form, such as a car, could be copied, rotated, and then merged with the other half). You could click the appropriate manipulator, grab the object with your mouse, and flip it around. Even watching the numerical readout on your screen, however, you would find it difficult with a mouse or pointer to hit the 180-degree point exactly. Using the numeric input, however, you can quickly click into the appropriate window to define the axis or rotation, type in **180**, click OK, and the object pops into position exactly as needed.

Extrusion One way to create a 3-D object is to create a polygon in two dimensions (for instance, on the x-y plane), enter the orthogonal window that is perpendicular to the original (the x-z view, for instance), and *extrude* the polygon (along the z-axis in this example). This is a useful way to create complex objects that cannot be obtained by using primitives. Extrusions can also be executed on lines (to form a plane) or on a group of lines that defines angles and curves. The results will never be fully formed objects, but the results can be used to create panels (which can appear as objects from certain views) or portions of an object that are assembled in a later stage by merging the panels.

Figure 5-3 shows a polygon extruded along a straight line on the z-axis. What would happen if the line were not straight or if the line deviated to some extent along the x- and y-axes? The effect would create a complex, snaking shape, such as that required to model the banister of a spiral staircase. Such an extrusion of 2-D polygons can be executed using special tools, whereby a guiding line is drawn in three dimensions and the 2-D polygon is directed to extrude, using the line as a center point. Here the numeric input device is highly useful to specify the amount of subdivisions and

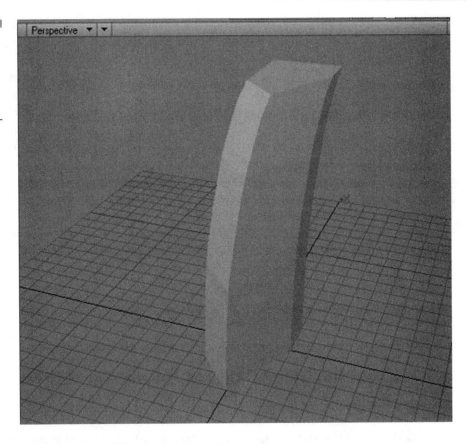

tapering of the 2-D shape, along the length of the guideline, which you may want to impart.

Rotation or Revolving   You can easily create interesting objects by rotating a closed polygon or curve around a defined point or axis. For instance, if you begin with a half-circle curve and rotate it around the center of the curve, you will create a sphere. If you take a 2-D circle and revolve it around a point or axis outside the circle, you will get a donut. Like extrusions, rotations and revolutions are useful tools to automate the process of creating shapes. Always consider these options when beginning each element in the labors of modeling.

Boolean Tools   Maybe you remember the term **Boolean** from your high school math classes. Boolean algebra uses a limited vocabulary of terms, such as AND, OR, NOT, and NOR to define interactions of groups. For instance, a Boolean command of AND (boys AND girls) would result in the joining of two groups, whereas the command NOT would result in a subtraction of groups (Americans, NOT New Yorkers). Animation programs enable Boolean operations on objects by allowing operations built on Boolean terminology.

For instance, you may want to join two spheres with a cylinder to form a barbell object. To do so, you would place the two spheres at an appropriate distance and add the cylinder between them.[2] You would then initiate a Boolean AND (also called an ADD) function, and the three objects would be merged. This seems rather simple, right? But how about the command, NOT (or SUBTRACT)? This function is much more fascinating.

Let's suppose you wanted to take a sphere and drill a nice, neat hole through it? Start with the sphere and superimpose the cylinder so that it passes through the sphere at any angle. Then use the Boolean command NOT (or SUBTRACT) and assign the cylinder as the subtractor. Bingo. The cylinder disappears and a hole in the sphere, where the cylinder was, is created. Using Boolean functions can provide a wide range of complex objects.

For instance, you might spend hours digitally sculpting the outline of a perfect ear, but an ear has thickness and you need to create the inside of the ear. You could start with a solid shape, duplicate it, shrink the duplicate a bit, and then execute a Boolean subtraction of the smaller ear from the larger ear to produce a perfect interior sculpting of the larger object.

Splines

After several years of working with polygons (during which a billion or so "flying logos" and other animations were made, based on boxes), the designers of animation programs began to look for other ways to define surfaces and model objects. Realizing the shortcomings of polygonal modeling, the designers searched for a way of defining surfaces with mathematically defined curves.

In this process, the designers realized that the computer used polygons to define specific sets of points in 3-D space. If these points could be expressed instead by formulas, the resulting shape would be smoother, easier to calculate, and more flexible.

---

[2]Different programs handle this function in different ways. Some enable the object to be superimposed in one scene, whereas others provide separate layers. Consult your program's Boolean instructions to determine the best way to practice this function.

The following explanation has a lot of underlying math that is really complex, the kind of stuff that could make a creative artist yawn, drop the book, and take a good long snooze. But this stuff is important to understand, so I'm going to show it to you in easy steps. Once we're done, you can take that snooze, but I think you'll want to jump back into the tutorial CD and try out some of the things you've learned. Knowing this material is what will make you one of the stand-out people in an animation studio whenever one of the new guys who never learned the basics of modeling asks a dumb question like "What's the difference between polygon modeling and **Nonuniform Rational B-splines** (**NURBS**)?" You'll know. You'll explain. You'll be the Big Kahuna.

Let's consider a typical animation task, making a sphere into a cartoon head. Using polygon modeling, we might grab several points and pull them away from the center using various tools. Eventually, with trial and error, we would create something that looks like Figure 5-4.

Give this exercise a try in any of the three programs, taking a sphere primitive and using polygon-modeling tools. You'll soon see how exhausting the process becomes. Eventually, you can get good at it, but a better way exists: using **splines**.

Figure 5-4
A sphere expressed in polygons, with several points drawn away from the center to form a nose

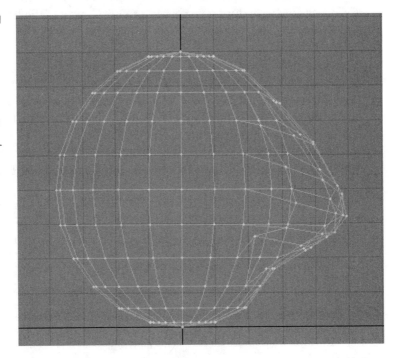

Olaf's Invention   The Vikings were probably the inventors of the spline. They hated sailing the oceans in wooden boxes, so this guy, Olaf, sat down to think of how he could make wood curve. First, he took three wooden posts and pounded them into the ground to form a straight line. Then he took a plank of wood and angled it between the first and second post. Now came the hard part. Heaving with all his muscles, Olaf bent the board around the center post and back behind the third post. Presto! He accomplished the first step of building a curved boat. Of course, when his wife, Nurbsilda, came and yanked out the posts, the board flipped out and whacked Olaf in the face, but that's another story.

The important thing we're showing is that points (mathematically known as vertices) can be used to *define curves* by acting as controls over a line. These points are called ***control vertices*** (**CVs**). The resulting curved line is called a **spline** (see Figure 5-5).

If we stack splines, like tic-tac-toe, crosswise over a 3-D space, we can define polygons. The CVs would then hover over this grid. By moving the CVs, we could control the shape of the form beneath. In the same way we defined **meshes** for the polygon-modeling method, we can define the array of CVs as a **spline cage** (many animators call this a **control cage**). The polygons that are formed within the cage are called **patches**. Patches,

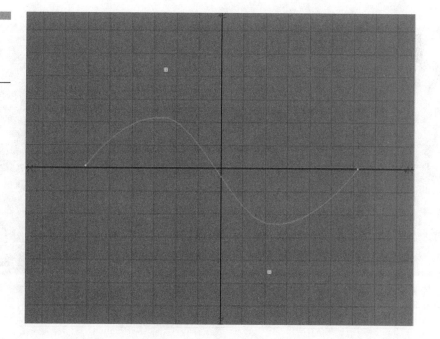

Figure 5-5
A simple spline curve with its controlling CVs

unlike polygons, can be twisted in 3-D space during the modeling process. You do not need to subdivide patches in order to render the surfaces. The computer does this exhaustive step automatically as part of the render process. This process is called **tessellation**.

Although an operator never needs to see the process of tessellation, 3-D programs can bring it to view by converting a spline object into a polygon project and vice versa. LightWave is one product that makes use of the reverse process in its proprietary MetaNURBS function. LightWave users can model with crude, blocky polygons and then execute a MetaNURBS conversion to convert the object into a spline-based object. More on NURBS in a moment.

You should also be familiar with how the computer measures distances on spline meshes. Because the Cartesian plane is already using x, y, and z, the splines use other letters: u and v. These two letters represent the equivalent of x and y positions on a flat Cartesian plane. The third direction, up and out from the surface, similar to the z direction on the Cartesian plane, is called the **normal**. Occasionally, you will hear reference to this form of spline measurement as **UV values**. Now you know what they are.

Spline modeling greatly reduces the amount of points needed to define and edit a 3-D object. This makes them the ideal method of modeling when you want to create gracefully curved objects that require intricate deformations, such as a character (see Figure 5-6).

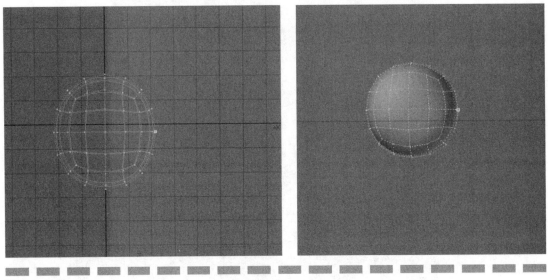

Figure 5-6
A sphere composed of splines and a cage of controlling vertices surrounds the sphere.

Types of Splines In your exploration of 3-D animation, you will encounter many types of splines. In 2-D graphic programs, a common type of spline is the Bezier curve. A CV on the line controls a Bezier curve and two attached handles can be manipulated to cause the curve to bend, kind of like a magnetic pull, with various positions of the handle (see Figure 5-7).

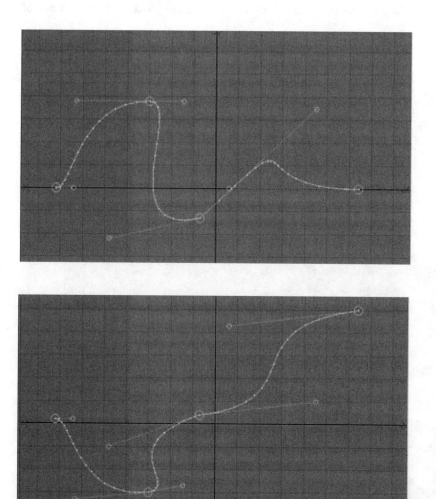

Figure 5-7
Various effects can be created with Bezier curves by pulling the control handles out and angling them in different directions. Nevertheless, the process is laborious and not always predictable.

A further development on the Bezier curve is the B-spline, which moves the CV off the line and allows for easier control of the curve (see Figures 5-8 and 5-9). Bezier and B-spline curves have one serious drawback, however, for use in modeling.

Olaf's Wife Saves the Day You remember our Viking, Olaf, and how he got whacked in the face when his wife Nurbsilda yanked the posts out of

Figure 5-8
Example of a B-spline curve with CVs effecting a predictable deformation.

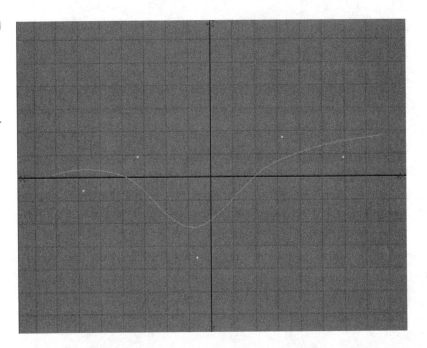

Figure 5-9
A deformation is executed on the spline sphere in Figure 5-6 by moving one or more of the CVs farther from the center.

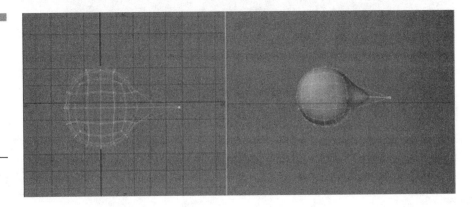

the ground? Well, Nurbsilda felt really sorry, so she thought a bit and, remembering how her spaghetti (her folks were from Liguria) got soft in hot water, she came up with an idea. She took Olaf's boards and softened them in boiling water. Then, when Olaf bent them, they stayed bent. Even better, Olaf could cut the board anywhere on its length and the board would still stay bent exactly as before.

Nonuniform Rational B-spline (NURBS) Modeling

Olaf had the same problem you do if you use Beziers and B-splines, because these kinds of splines cannot be cut at any point in the curve and then easily pasted into another curve. Creating complex objects such as the ears of a donkey or the wings of a dragon would be difficult using splines that could not be cut and pasted at will (see Figure 5-10). Although the reasons for this limitation are too technical for this book, you may be happy to know that a further development in curves came with a type of curve called a NURBS.[3]

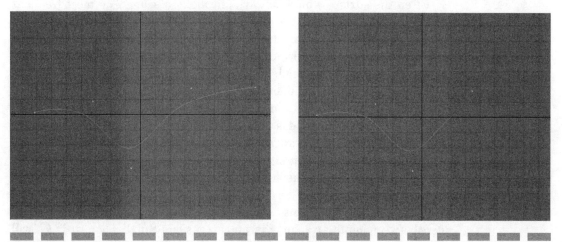

Figure 5-10
An example of a key benefit of NURBS. The left illustration shows a typical NURBS. If this were a B-spline, cutting it would most probably produce unpredictable results in the curvature, but a NURBS can be cut anywhere while preserving the original properties.

[3]Keep in mind that the term, NURBS, is an acronym. Therefore, it is always written capitalized and although the word ends with an "S," it is a singular noun.

I'd like you to believe that NURBS was named after Olaf's wife, Nurbsilda, but I just made all that up to help you remember the concept.

Keep in mind that although many types of modeling are available, each type boils down to meshes of polygons. A computer cannot render anything but polygons. No matter how smooth a rendered surface may appear to be, somewhere down in the tiny world, straight-edged polygons form your object. Shading creates the smoothness. Later, when we discuss surface attributes, I will show you some different kinds of shading that will yield spectacular results when you render any polygon mesh.

I bet you think you're all done with modeling, yet as that annoying guy with the $19.99 Ginsu blade says, "But that's not all!" No siree. The technology of modeling goes on and on, developing further features and benefits.

Until now, we've been discussing methods of modeling that create surfaces or skins, resulting in vast meshes of polygons. All of these make great static objects, especially NURBS. The success of these methods has enabled animators to concentrate on their primary desire, making characters move as good as Disney cell animation. Actually, they weren't all trying to out-Disney Disney, but that's considered the pinnacle of many computer animators' dream pyramid. So what happened?

Often the surface would simply split along a seam. The most frequent place was the shoulder of a human figure when the arm went from out straight to down flat along the hips. Ugly black holes would show on the shoulder of the beautiful princess just as the prince asked her to dance. "My dear, but you have a gaping black hole in your shoulder!" he'd say.

Clearly, something had to be done, so the magicians who write animation program code developed a kind of marriage between NURBS and polygon modeling. This new method is called **subdivision surfaces** (AKA **patches**) and it has two significant advantages over anything that came before.

First, subdivision surfaces are a solid modeling system. In other words, the polygons have no backsides, and the computer considers the interior of the shape to be filled. Therefore, holes can't show. Second, because no NURBS are used, no seams can be split open. All the splines cross all borders between joints, attachments, and extensions. Subdivision surfaces still employ a control cage of CV points and still employ splines, so all the familiar tools that animators learned in previous modeling methods are preserved. Those animators who've been around since the days when polygon modeling ruled are happy too, because subdivision surfaces constantly display the underlying polygon structure (this is called **polygon proxy** mode), allowing each polygon to be addressed easily.

Incidentally, from a historic point of view, each of our animation programs owes its origins to polygon modeling, so polygons are still featured.

Softimage and Alias Power Animator, the ancestor of Maya, first developed NURBS. LightWave, at the lowest end of the cost scale, never seriously developed a NURBS capability but instead developed a simple utility for smoothing polygon forms in a way that was similar to NURBS, which they installed as MetaNURBS. MetaNURBS is really a means of subdividing polygon surfaces, and its development pushed the higher-priced programs to advance the science of subdivision surfaces. Eventually, both 3ds max and Maya spent considerable time adding NURBS to their polygon toolset and then added subdivision surfaces so that today these two programs offer modeling in all three methods.

Now I hear you exclaiming, "Okay, darn it! I'll buy that Ginsu knife and the free Chop-o-Matic. Can I take that snooze now?" No, there's still more! Here's a really clever advancement that I know you'll love. Let's say you have a human figure named Vernon Salvage, created with NURBS or polygon modeling, and it has 8.6 gazillion polygons. Now the reason you made all those polygons is because in some scenes Vern is going to be close to the camera. We're going to see his tattoos and his pierced nostril. So we need lots of detail, right? We need those polygons, baby!

But in other scenes, we're going to show Vern in long shots on the beach, Big Sur pounding on the sand and white clouds in the sky. In these shots, a dozen or so polygons will do the trick to form Vern. What would you do? Design two models of Mr. Salvage? You'll also need a series of medium shots and and a two-shot with his girlfriend, Penny Nicoletti. Thus, lots of versions of Vern at different levels of detail will be needed, which means lots of work if you're going to make a different model for each. It also means lots of rendering if you're going to have the computer render 8.6 gazillion pollies, when all you need are 12. That's why you need LOD, Sam!

Level of Detail (LOD, AKA Parametric Modeling)

LOD is a feature of certain subdivision surface modeling systems and it actually adjusts the amount of polygons proportional to the closeness of the camera. This is incredible when you think about it. Zoom in—the polygons multiply to remain at a density that supports the detail you require. Zoom out—the polygons reduce to increase the render speed. Different forms of LOD are available, often as plug-ins, for nearly all professional animation programs.

Okay, folks. You've bought the Ginsu, the Chop-o-Matic, and the 24 audio tapes on how to buy real estate with no money down. We're done with modeling. Obviously, in a fundamental book such as this, I have left out lots of technical details. I hope you have a good overall scope of the different kinds of modeling, how they've developed, and what each can do best. If you want more information, you can delve deeper into the programs on the CD, buy books that describe each of the three programs in much more detail, and maybe buy one of the programs and start modeling for real.

Object Terminology

When discussing objects, it's helpful to know about wireframes and control points.

Wireframes   Until now, we have been discussing objects in terms of points and polygons. These same objects can be displayed on your computer as **wireframes**. Wireframes are displays of objects as defined by the points and connecting lines, with no surfaces defined (see Figure 5-11).

When computers were not as powerful as they are today, wireframes were all an animator had to display his objects until the final rendering. Can you imagine showing a client a preview composed entirely of wireframe objects and scenes? This is what we did back in the '80s and the early '90s. Today computers with **OpenGL** video cards can easily display shaded objects and even impart some degree of **surface attributes**. Wireframes, however, are used during most of the modeling phase.

After an object is created pretty much to your desires, it is time to define the surfaces. A surface is a polygon defined by any number of points and assigned with a range of attributes. Without surfaces, the points and lines

Figure 5-11
A complex object displayed as a wireframe and as a shaded solid

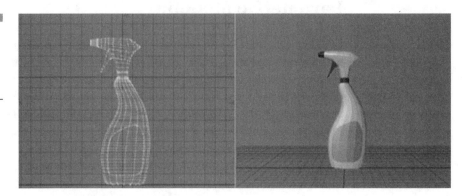

of your object designs would not be visible. It is the surfaces by which the computer literally creates your objects.

It is possible to define and render distinct attributes for every polygon on every object, although this is rarely done. However, it is common for a complex object, such as a human face, to be composed of several groups of surfaces, each with different attributes. For example, the lips of a woman's face would have a different color and reflectivity than the rest of her face.

When **surface modeling** (as opposed to **solid modeling**) polygons of more than two points, the surface will obviously have two faces (front and back or inside and outside). Animation programs enable the user to define different attributes for each face. Commonly, a surface-modeled object will be composed of a closed surface. A cube, for example, would have six faces facing outward and six facing inward. If the inside of the cube is never shown, the animator could eliminate any attributes for the interior surfaces and make the object less complex to render.

For this reason, animation programs enable us to assign **surface normals** to each polygon. Surface normals are usually displayed as a faint line sticking out at right angles from the surface and pointing in the direction the surface will render. In our cube example, for instance, six surface normals would point out from the cube. Because the inside of the cube would not need to be rendered, no normals would point inward.

Animation programs offer utilities for **flipping normals,** should the animator want to make changes. They also provide a means for **doubling normals** to create polygons that will be rendered on both sides, although this effect is rarely used.

Control Points   Once an object is created, its destiny is to be placed into a 3-D environment, positioned, and perhaps moved. Animators call this **staging the object**. Control points and hierarchies are two factors that can be associated with an object during construction or staging.

Control points, also known as **pivot points**, are unseen points around which the object will turn whenever it is moved. For instance, the best pivot point for a spaceship might be in its geographic center, whereas the best pivot point for a door will be along the doorframe where the hinges are mounted.

Most programs offer the capability to place a pivot point on an object during creation and then to adjust the placement of that point during the staging and motion-scripting stage of the animation process. Although most well-known objects, such as spaceships, doors, and cars, have expected positions for their control points, objects you create from your imagination, such as logos, have control points anywhere inside or outside their perimeters.

Experimenting with the placement of control points can reveal many interesting motion effects for imaginary objects and bring new dimensions to your work.

The Relationship of Light and Objects

In the real world, our eyes see objects because light is bouncing off those objects and coming into our eyes. Light from the sun and most artificial sources is considered white light. In reality, all light has some color to it.[4] In this chapter, we'll pretend all light is white, unless otherwise stated. As you may know, white light is not just one color; it is the combination of all colors. You can prove this by putting a glass prism in a white light beam and breaking the light up into its full spectrum of color **wavelengths**. Each color has a specific wave frequency that defines the color. So if all light is white, how come we see bricks as red and blueberries as blue?

Objects in the real world appear to have color because their surfaces, at the molecular level, are absorbing various wavelengths of light. In other words, a brick is red because it is *absorbing* all the blue, yellow, green, and chartreuse light wavelengths and *reflecting* only the reds. A blueberry is blue, because the blueberry is absorbing most of the red light that hits it, so it doesn't look like a strawberry.

Now you know why real objects have color. What about texture? Why does a new car on the dealer's lot look shiny, whereas my car, parked at the train station with six days of winter salt and road dirt on it, looks dull and flat? And don't say it's because I'm too lazy to wash it. It looks flat because it is reflecting less light than the shiny new car. Call my wife and tell her that, would you?

The appearance of all objects in the real world is dependent on their surface's **attributes** or **parameters** in relation to **reflecting light**. In our artificial world of computer animation, the same rules apply. They may not be as complex as the real world, but we're getting closer ever year. Every

---

[4]You can easily see the color of white sunlight as opposed to light from a white lightbulb when you take a picture indoors with indoor light and have sunlight coming in through a window. The window light always looks bluer than the interior light. These differences in white light are due to color temperature and are measured on a **Kelvin** scale in degrees. Sunlight, for instance, is 5,200 degrees Kelvin, whereas lightbulb light is about 3,600 degrees Kelvin. These differences appear in motion picture film and can be employed by sophisticated animators who want to model the color responsiveness of film either to imitate a filmed scene or to composite real-world objects with *computer-generated images* (CGIs).

surface of every object in your animation will be seen because you impart upon it some quality or attribute that affects how light (supplied by the computer) will reflect off that surface.

Shading Remember when we were working with modeling objects how we started with the earliest solution (polygon modeling) and worked through each development that was still in use? How about if we do the same with shading? (Oh no! Not another Ginsu!)

Shading is the process by which your computer displays and renders the polygons that make up your objects. Keep in mind, as noted before, that what you see onscreen as you work (the product of your video card) is only a representation of what the computer will render (the product of your **central processing unit [CPU]**) for the final sequence of pictures that will make up your animation. Each version of what you see, however, involves shading, but your computer, working for minutes or hours per frame, can do a hell of a lot better shading a polygon than the OpenGL **application programming interface (API)** of your video card, which works in microseconds.

The most primitive form of shading in computer animation is **constant shading**, also known as **flat** or **faceted shading**. This kind of shading enables only one color to be displayed across the surface of the polygon. Constant shading, as the name implies, renders the polygon flat, with sharp edges and vertices. You've seen constant shading at work in older, cheesier video games. Almost no one uses this shading method anymore, unless they are purposely trying to make something look retro or extremely faceted.

The next development in shading came about when animators starting making greater demands of curved surfaces. Just like polygon modeling developed the capability to model curves by dividing the polygons finer and finer, shading followed suit with **polygon mesh shading**

Polygon mesh shading offered some interesting developments, such as the capability to display a range of colors across the polygon's face. The idea was that roundness could be imitated by creating small highlights. At the edges of each polygon, however, the effect broke down, and due to the physiology of the human eye, there never seemed to be a point of polygon subdivision where the faceted effect disappeared. Our eyes were simply too good to be fooled. Because each polygon was shaded individually, and because each polygon was angled slightly different from its neighbor, an ugly effect always occurred at the joints.

Then, in rapid succession, came two new ways of shading that allowed each polygon to be shaded in direct proportion to its neighboring polygons. These methods were **Gouraud** (pronounced "goo-ROW") shading and

Phong shading. A few years back, computer animation programs offered both forms of shading because Gouraud was a bit faster and Phong was a bit better in quality. Today, with computational speed high enough to use the most sophisticated forms of shading, Phong shading and an even more accurate method called **Blinn shading** are the included default ways most animation programs render polygons.

All this talk about shading so far has been concerned with smoothing the sharp edges of polygons where they meet. But what about edges that you *want* to remain sharp? Let's say the cutting edge of that Ginsu knife that is otherwise smooth and rounded along the blade must stay sharp. For situations like this, you need a method to tell the computer which edges you don't want smoothed by Mr. Phong or Mr. Blinn. Some programs offer a limiter, based on the angle (of the **surface normal**) formed by any two adjacent polygons. For instance, if you want every angle less than 90 degrees to be smoothed, you could set a **maximum smoothing angle** to be 89.5 degrees. If the computer sees anything less than that, the edge is smoothed. If the edges meet at 90 degrees or more, the edge is left as is. Smoothing parameters can save you the effort of creating two or more smoothed surfaces and then joining them along the sharp edge, which would be the hard way of making that darned Ginsu blade.

As computer science further developed, creative individuals began to offer new methods of shading, now generically called **shaders**, for a wide variety of specific purposes. Some shaders are named after their creators, such as Strauss (good for metals) and Oren Nayar Blinn (good for creating fuzzy effects), whereas others are named for what they do best. **Anisotropic** renders a kind of brushed metallic reflection, and **shag fur shaders** make a surface look like shaggy fur.

Many shaders are now available for LightWave, Maya, and 3ds max from third-party vendors as **plug-ins**. For instance, **Darktree Shaders** (www. darksim.com) make excellent plug-ins for 3ds max, LightWave, C4D, Hash, and True Space animation programs that enable you to zoom in on an object. Darktree Shaders are yours, free on the CD I've included in this book! Because the shader recalculates its parameters at all focal points, large texture maps are not needed, and they tend to slow the rendering anyway.

Finally, because each of the key animation programs offer some level of open architecture, which means you can write your own code for certain areas of the program, many animators with programming experience write their own shaders.

Advanced Shading Techniques When the computer starts to render a surface, it examines the tables of information that have been assembled by the animation program and that are derived from your inputs in the control screens. Creative programmers have examined the process by which the surface is generated and have introduced certain software procedures that enable unusual effects to be created without a great deal of labor on the part of the operator.

Take, for instance, the issue of hair on a character's head. In the past, a single hair could be expressed in some 3-D computer programs as a cylinder with a small diameter and in others as a two-point polygon. Such a strand would have to be duplicated hundreds, if not thousands, of times to cover a human's head, and even more to cover the skin of an animal figure. Such hairs would remain fixed, either straight like needles or bent in some fashion, unless some keyframable deformation could be expressed on each. Later, you'll see how we accomplish this with arms and legs, but hair? Millions of hairs? I don't think so.

You can imagine the complexity that hair can represent if each hair must be created as a single object. This problem challenged many animators before you. Then some brilliant programmers came up with the idea of dealing with hair as a shader issue.

Instead of creating an individual pixel at a particular point on the surface of an object, the computer creates a strand of hair at predetermined intervals on the surface pixels. This strand, like all other strands emanating from the object, will have a root point (the same as the pixel the strand replaces), a set length (determined by the operator), and certain flow properties. Flow properties might include the original direction, stiffness (or resistance to movement of the underlying skin), and memory (or the tendency to return to the same original direction). See Figure 5-12.

Shaders are currently considered one of the most advanced levels of computer animation and may not be found in all animation programs. Look for these features as optional plug-ins that may be offered by third-party vendors for your animation program. Other animation programs enable an experienced software engineer to write code that will work as a procedural or shader effect.

Surface Names Now that we can shade a surface well enough to render realistic curves, what are the attributes that you might want to impart on those curved surfaces? The first and most important attribute of a surface is its name. As mentioned before, an object can have many surfaces

Figure 5-12
Examples of hair
shading using a
third-party plug-in
from Digimation's
Shag: Hair ($495)
for 3ds max.
Knight image,
copyright 1999 by
MagicPictures,
image created by
Makoto Chiba.
Eeevil Guy,
copyright 1996–
2003 Digimation,
Inc. All rights
reserved.

composed of one or more polygons that serve a distinct purpose on the
object. They could be the shiny red skin of an apple, the stem, or a leaf on
the stem. All these surfaces, if they are going to have attributes or **parameters**, need to have distinct names so that you and the computer can recognize them, set their properties, and render them correctly.

Each animation program has a means by which polygons can be selected and grouped into a surface and named. A surface is usually named after its object and then by its specific purpose. Then if several surfaces have identical purposes, they are numbered, such as TableLeftFrontLegface2. Of course, if all the surfaces on an object have the same parameters, you needn't waste time giving them separate names.

Parameters of Shading   We've already brought up the attributes of **color** and **shininess**. Shininess, by the way, is called **specularity**. The shinier an object is, the smaller the highlight that will reflect from it. Specularity is related to another attribute, **glossiness**. Glossiness determines how the highlight spreads out from its center to the end of its effectiveness.

A good way to remember the difference is to think of a used car lot. The cars are often lit by lots of small lightbulbs to make the cars look shinier. The size of the highlights on the surface of the car is the specularity. The sharpness or dullness of the highlights is glossiness, and whether you see the bulbs or just the light is defined by the reflectivity.

Reflectivity is the value that determines how much of the surrounding area will be reflected back by the object. In the real world, shiny objects, such as mirrors, cue balls, car fenders, and water, reflect the world around them. Even a simple metallic logo hanging in the center on your TV screen can be made twice as stunning if it reflects a mountainous desert plane from somewhere off camera. Most animation programs offer some way of giving objects reflective properties.

For instance, if you were looking into a mirror and if the reflectivity were set at maximum and the transparency set at minimum, you would see everything the mirror faced. A windowpane might have a little less reflectivity, but a lot more transparency. Each attribute is usually expressed in numeric values, most often in ranges of 0 to 100 percent. An object of zero reflectivity, glossiness, and specularity might be a piece of velvet.

Although the controls for achieving reflectivity are rather simple, a lot of mathematically complex work goes on in your computer whenever you add reflectivity. It requires the computer to calculate the surface properties of the object, the presence of other objects, their reflectivity, the presence of light rays, and the presence of image planes that act like large pictures hanging in the scene.

The computer takes on all of these factors one at a time, sometimes ray by ray, and calculates them into a finished frame, employing a much higher degree of reality than any image without reflections could convey. This is called **ray tracing** and takes lots of work. Animation programs have some tricks that mimic real ray tracing for those times when you want the effect but don't have the time.

For instance, it's possible to have zero reflectivity and high specularity. The object's surface would have pin lights reflected, but it would reflect no surrounding objects. This is a good way to make something shiny without causing a lot of rendering time.

An object can be assigned varying levels of reflectivity from zero to 100 percent (see Figure 5-13). Zero reflectivity would result in a dull, matte object, such as velvet or unfinished wood. High reflectivity would be required for shiny metals or lacquered surfaces. The highest reflectivity would be a mirror that returns all light and reproduces the images of nearby objects.

Diffusion is another attribute relating to light reflection. Highly diffuse objects tend to scatter the light that falls on them. Sandpaper is a good example. What would happen if you shined a really bright light on a sheet of sandpaper? That rare event would be the combination of high specularity and high diffusion, which would kind of work at cross purposes.

Refraction is a surface attribute that can be used on transparent objects to create internal depth where none really exists. Imagine a pencil stuck into a glass half filled with water. When viewed through the glass, the pencil seems to break at an angle below the water's surface. You remember your eighth-grade science teacher telling you this was called refraction, an effect caused by the fact that light is slightly bent by the water, right? But how would you do this effect in computer animation? Two ways.

Figure 5-13
Two different renderings of a scene containing a knife blade set with reflectivity off (left) and on (right). Which seems more real to you?

Time-Saving Tip

In Figure 5-13, the attractive reflection on the knife blade was created by ray tracing, and even on my IBM T31 laptop where I made these images, the effect took only six seconds to render. But what if you had the knife blade reflecting an entire city. Would you have to render each building? That could take weeks!

In such cases, you could opt to acquire a digital photograph of a city and place that somewhere in the scene to reflect off the knife blade. In the old days, the only way to do this was to create a flat surface, map the photo on that surface, and place the surface beyond sight of the camera and angled to reflect off the reflective object. This method could still be used if you had a lot of objects with different reflective parameters and you wanted to quickly impart on all of them the same image.

However, if you have only one object into which you want to reflect an image, animation programs offer a variety of attribute tools that enable any surface in a scene to be mapped with an image. These are called **reflection maps** (see Figure 5-14).

Figure 5-14
The surface of the knife blade has simply been given a reflection map image of Hong Kong to create the impression that the city is located somewhere behind the camera.

You could create a water object, make it denser than the surrounding scene, and execute a complex ray tracing to get the virtual light in your scene to bend the pencil. You could also use the refraction surface attribute, which does a pretty good job of pretending all of the above. Then all you'd need to do is name the surface of the glass below the water line differently than the top of the glass, put a horizontal water surface at the dividing line, and add some refraction to the lower surface of the glass. You don't use this attribute much, and maybe I've spent more words here than you think necessary, but this is one powerful example of how surface attributes can make your work easier and your scenes faster to render. And that translates into money in your pocket.

Transparency is, obviously, how invisible the surface is. This attribute would be a careful consideration when creating a pane of glass. Transparency is similar to **translucence**, but not the same. Translucence takes into consideration the tendency of a transparent object to diffuse the light that travels through it. A good example of translucence would be a pane of glass that is thick, or perhaps somewhat foggy or frosted.

Some programs address the transparency of an object's edges separately from the entire object. In real life, sharp edges of objects, particularly objects that are somewhat transparent or highly reflective, look more opaque at their edges. If the light is passing through an object, this phenomenon is known as **refraction**. If the light is bouncing off an object, the term is **reflection**.

LightWave addresses this issue with a separate control, **Edge Opacity**. 3ds max enables the entire object to be assigned an **Opacity Falloff,** which allows the transparency of the object to vary from the edge to the interior in either of two directions (**in** or **out,** as shown in Figure 5-15). Maya takes a more real-world approach, enabling the user to define the **refractive** or **reflective index** of a ray-traced object.

Luminosity or **self-illumination** adds an internal glow to the object. Imagine a yellow bug-light bulb, plugged into a dimmer switch over the corner of a room. The bulb represents a colored object in 3-D space. Its surface is yellow, but if the dimmer is off and the room is dark, the bulb is black. If we turn the dimmer up slowly, the bulb begins to illuminate from within, giving off yellow light. In a computer animation program, the amount of light coming from within the object is called luminosity or self-illumination.

Consider using luminance in sparsely lit scenes where they would otherwise tend to lose detail in darkness. It's also a good tool to use if you want to limit the amount of light instruments in the scene (more on lights in Chapter 6). Luminosity can also be used to create a backlit effect, as if an object were made of translucent plastic with a lighting source within.

Figure 5-15
The Material Editor in 3ds max enables parameters to be set for a glass tube object running across the foreground. Notice under Extended Parameters/ Advanced Transparency, the amount of falloff is set at 75, the direction is "In," and the Index of Refraction is set at 1.5. The results are shown in the OpenGL Perspective view (lower right) and the test-rendered frame (upper right). Notice how the edge of the tube is less transparent than the center, as it would be in real life.

Luminosity affects only the selected object and is used to light other objects. Luminance is therefore not suitable for creating light-emitting objects, such as the red blinking light on a police car, where the red light would be cast on the surrounding scene. Luminosity can be used for objects such as lightbulbs, lanterns, neon tubes, and the red-hot poker in that masked man's knurly fist. Other luminance attributes include **glow** and **incandescence**.

Consider a cylindrical bar of color with some self-illumination, such as a neon tube. If the bar has a high degree of edge opacity, the bar would be well defined against its background. If the bar has a lower degree of edge opacity, the bar would appear fuzzy, as if seen through a fog. At the lowest level of opacity, the bar would actually cease to exist and the object would appear more like a ray of plasma.

Some programs provide colored-edge opacity or the outlining of an object. When applied to a somewhat transparent object, a colored-edge outline can give the appearance of a certain thickness to the object's skin, suggesting, for instance, the inner surface of a glass.

Comprehensive Object Effects   Up to this point, we have been describing individual surface attributes and the collection of tools you can use to achieve realistic effects. We now move on to more comprehensive effects that have a great impact on the illusion of reality. Although the following tools can be applied to a single polygon or an entire object, they tend to be used by animators in a comprehensive manner. Therefore, I classify them as comprehensive object effects. This sounds sophisticated, but actually it's just an easy way to group them. The following effects are the most powerful and most frequently used in all three animation programs.

Color is usually selected from a set of sliders, a color wheel, a hexagon, or a grid representing a system of color definition. The two most popular systems are **RGB**, for *Red-Green-Blue,* and various levels of ***hue, saturation, and brightness*** (**HSB**).

RGB represents the earliest form of color control where a range of numbers represents each color element's intensity. Originally, the color range was from 0 to 15 for a 16-bit color system, yielding a maximum of 4,096 colors ($16 \times 16 \times 16$). Today the range of numbers is commonly 0 to 255, yielding a total of 16,777,216 colors, more than enough for the human eye.

RGB is called a subtractive color system. This means that as each color is added, the resultant mix of colors moves toward white, or the absence of color. Therefore, the maximum amount of each color (255, 255, 255) is white, whereas the least amount of color (0, 0, 0) is black. Mixing maximum amounts of red and green, on the other hand, (255, 255, 0) creates a maximum brightness of yellow.

It is important to note that the limit of the RGB system, or **gamut**, is greater than the signal capacity of NTSC television. Settings above 220 RGB may cause colors in NTSC to smear past their intended borders.

HSB handles color mixing in a different fashion. This mixing is usually offered on a color wheel with black in the center, white around the edges, and various colors appearing as pie-shaped slices around the circle. Hue refers to the actual color definition, which, on a circle, would be represented by the amount of degrees from the 12 o'clock position. Zero degrees is red. Sliders for each parameter are also included for those who are not circularly inclined (see Figure 5-16).

Saturation represents the amount of color between black (0) and full color (255). Brightness represents the amount of white in the color, which sounds a bit like the opposite of the definition for saturation, but it is not. Zero saturation means no color, whereas zero brightness means the darkest form of the color.

Materials is a collective term applied by 3ds max and Maya to apply to the entire set of information that describes how a surface will render. Light-

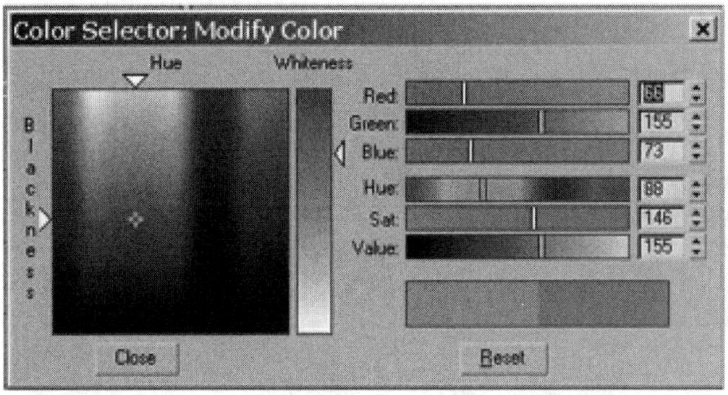

Figure 5-16
The color selector for 3ds max. Two swatches over the Reset button represent the original color on the left and the new color being made on the right. Moving any of the sliders above the swatches or triangles around the spectrum displays moves the cursor on the spectrum and vice versa. When the new color is selected, the user presses Close. Numeric entries can also be made in the rightmost column.

Wave calls the set a *Surface Editor*. Because LightWave separates its program into Layout and Modeler (where all surface definitions are set), the Surface Editor is found in the Modeler portion of LightWave.

Materials in 3ds max and Maya also refer to a collection of surface sets displayed as an array of small rendered icons. An entire set of surface attributes is stored in each icon. This paradigm of design offers a cyclical path of use. For instance, after a surface is created or selected, the operator goes to the materials array and clicks the appropriate icon. The selected surface instantly acquires all the attributes within the material's set. The Material Editor window can be opened to display the entire vast array of surface attributes, some of which have been listed previously (see Figure 5-17). If these attributes are changed, the icon in the materials display changes quickly (because it is small and simple, little or no render time takes place). The new material can then be saved as a newly named icon and used again in the future.

Although LightWave does not offer an array of materials for quick selection, the Surface Editor displays one icon object that is quickly rendered as attributes are changed in the same way the selected icon in the materials array changes in 3ds max and Maya (see Figure 5-18).

Surface mapping is a technique that places either 2-D or 3-D images onto or into an object. We've already seen how a bitmapped photographic image can be projected onto the surface of an object using reflection mapping.

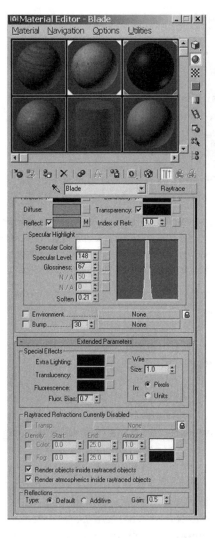

Figure 5-17
The Material
Editor in 3ds max.
The top part offers
an array of
interactive icons,
whereas the
lower portion
offers a vast array
of controls for
changing the
attributes of the
selected material.

Other bitmapped images can be projected onto the surface of an object to provide cylindrical labeling effects (such as a beer can), tiled effects (such as a chessboard), or spherical effects (such as a planet's continents and oceans).

Projection controls determine how the image will wrap on or around the object. A **planar projection**, for instance, places the image on a flat surface, such as a poster on a wall. A **cylindrical projection** wraps the image around a cylinder such as the label on a soda can, or around a consecutive series of flat sides, such as a label on the front and side labels of a cereal

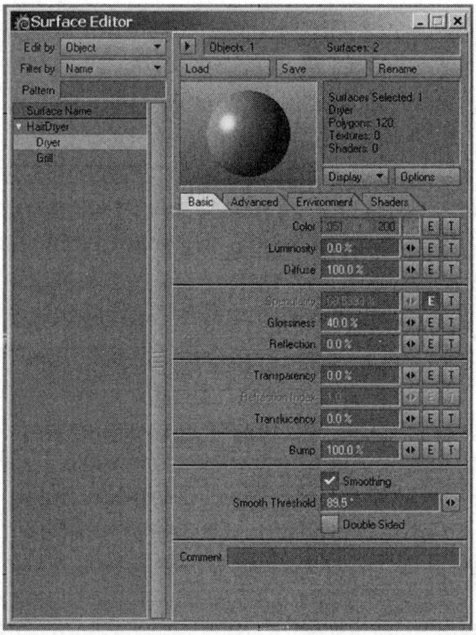

Figure 5-18
The Surface Editor in LightWave. The icon at the top will rerender quickly to show the changes made in the screens below. The surface attributes can be named, saved, and reloaded.

box. A **spherical projection** wraps the image around a sphere such as a map stretched around a globe or around all the sides and top of a cubic object.

In addition to photographs and static designs, the surface-mapping technique can also be employed with designs that the computer generates from random elements, designs we call **procedural.** Although pictorial maps represent a large photographic image that is usually equal to or larger than the object on which it is cast, procedural mapping represents images that are generated by the computer according to a mathematical formula and are usually much smaller than the host object.

Procedural surface mapping includes several simple categories that may be found in all animation programs. These include **tiling** (or **bricks**), which takes a small pattern and repeats it across the surface at programmable intervals. **Checkerboard** creates a pattern of alternating squares of two different colors. **Noise** creates a random pattern of any two colors with frequencies as small as the pixel elements. **Swirl** or **marble** imitates the serpentine patterns of marble, and **paint** allows you to use common drawing tools to paint the surface of an object.

Procedural surface mapping also includes a range of ever-developing effects categories that continually stretch the imagination of animators. Wood was one of the first of these categories and included most of the

Time-Saving Tip

Front- and rear-projected image maps are useful for quickly creating complex scenes. Imagine that you have a scene in which the camera will move very little or not at all. In this scene, you want to place a complex character that will take time to animate. You don't have time to render the entire background scene, so instead you project a photographic image. As long as the camera doesn't move too much, your **rear-projected image** can appear to be a full-developed 3-D environment: an office, forest, cityscape, or kitchen.

But what if you have to move the camera? In this case, you may still be able to take shortcuts using **image projection**. For instance, if you have a camera traveling horizontally along a road, the background elements will move horizontally in proportion to their distance from the camera. The distant mountains may not move at all. Relegate them to a static rear projection. The forest, just a few miles away, may move slightly. Make this on one front projection, with the sky cut away to show the mountains behind. Some trees might be in a field between the forest and the road. These can be individual **front-projection** planes that move horizontally in and out of the shot. Now, in the foreground of all of these projections, add your complex character running along, and the scene is complete in a fraction of the time you needed to make 3-D mountains, forest, and trees!

species on the planet. In quick succession came others: dirt, rust, waterdrops, rivets, veins, pestilence, you name it.

Third-party suppliers offer libraries of textures and images designed to create specific surface textures. Plug-in programs from companies such as Alien Skin Software (www.alienskin.com) enable artists to use popular programs such as Adobe After Effects and Photoshop to create textures and effects that can be exported into any 3-D animation system.

Soon artists began to develop their own textures and even copyrighted them. Ron Thornton, one of the great pioneers in 3-D animation and a master craftsman of space vehicles, once created a texture map for the surface of a grimy, old battle cruiser that became, for years, the de facto texture for a generation of space animators. Eventually, artists demanded the capability to paint directly on their objects in 3-D space.

Painting on the surface of a 3-D object is an effect included in some programs or offered as a third-party plug-in. Painting is extremely useful in creating organic models, such as character faces. In a 3-D painting program, the animator may choose any orthogonal view of the object, select a virtual brush or drawing tool, a painting medium, such as paint or spray, and a brush size. Using the cursor tool, the artist then applies the paint to the object and, in all views, the object changes appropriately to display the painted effect. This effect can be stored as a surface, unfolded, and used as a flat object or even copied onto other objects.

When procedural maps are used in 2-D mode, they appear only on the surface of an object. 3-D maps penetrate the interior of an object to enable higher degrees of creative flexibility. For instance, a 3-D procedural marble texture, when applied to an object such as a sphere, would enable the animator to cut a wedge out of the marble sphere and find that the surface's pattern was replicated appropriately in direction and mass beneath the surface.

Limits of Texture Maps　The term *texture* is a little misleading. I think of texture, in the real world, as a tactile attribute. When I think of the texture of a cantaloupe, I'm not thinking of how it looks, but how it feels in my hands. This might lead one to think that texture mapping in 3-D animation is a means of deforming the surface of the object. This is not so. Texture mapping only creates a somewhat randomized *photographic projection* on the surface of an object. The object is otherwise perfectly smooth or as smooth as the shading algorithm seems to make it! (My God, is nothing real here?)

Now consider this. What if we could use the data contained in a flat map (either a bitmapped photograph or a randomized procedural map) to interact with the formula that drives the shader? What if the shader, heretofore dedicated to rendering a smooth surface, could be perverted in its mission? What if we could contaminate the shader with procedurals? The result would be a bump or displacement map.

Bump Mapping or Contour Mapping　Bump mapping, also called contour mapping, is a means of performing elaborate deformations of the surface of an object without actually modifying the object beyond its basic shape. Bump mapping is an automated procedure whereby the computer takes your object's plain surface and processes it using a flat image map as a guide. The results can be quite startling.

When the computer executes a bump or contour mapping effect, it starts by wrapping the 2-D source image around the object, using many of the same placement and projection parameters as simple image mapping. Once positioned, the computer then examines each pixel of the map image in terms of its brightness. Pixels that correspond to bright areas on the map are rendered to appear at a higher level above the surface than pixels that correspond to darker areas. In other words, bright pixels are moved above the surface of the object, whereas darker pixels are given lesser altitude. A white spot, for instance, in the middle of a black area will, when bump mapped on a sphere, create a crater effect. A repeating series of small gray triangles on a dark surface might create a scaly, lizard skin.

The key to bump mapping is to supply a suitable black and white image that will produce the effect you want. A series of small rectangles, for instance, might be perfect for adding complex surface detail to a spaceship's external engine surface. A black and white photo of a piece of marble could produce an interesting sedimentary effect on the sides of a mountain object.

Keyframing Maps Projection maps and bump maps can easily be changed over time by using keyframes. A projection map of clouds, for instance, can be made to move over an object's surface to create a realistic scenic effect. When bump maps are gradually changed, incredible things happen. Consider a flat plane that is distorted by a bump map of swirling images. The result would be similar to the waves of an ocean. Combine two bump maps moving in perpendicular directions and you could easily emulate an ocean in a storm. Add a little lightning above a boat with a fearful crew and they might start praying for mercy.

Keep in mind that the result of bump mapping only happens in the rendering stage. It is not seen in most OpenGL or Active X previews, and it does not affect the actual surface geometry of the object. Bump mapping also has a rather shallow limit of effect. For deeper effects, you must use another technique, based on surface mapping, called **displacement mapping**. In displacement mapping, the actual geometry of a surface is changed by an image. This image can also be animated over the surface of the object, and if the image is a frame-by-frame sequence, the animation can be quite complex.

Displacement mapping is controlled in the same way as bump mapping, but in this procedure the amplitudes of light and dark in the source map's surface move the actual surface elements of the object. The changes become a permanent feature of the new object.

You've now been taken through all of the introductory levels of modeling and objects, although some of what you've learned may not be familiar to experts sitting at the next workstation.

Take some time now to catch up on your daily sales calls, practice the software on the CD, and then come back here for the "Lighting Lecture."

Let There Be (Rendered) Light

This chapter explains how light is used in an animation program to illuminate objects and scenes. It also shows how rendering assists lighting.

Theoretically, without light being defined in a scene, nothing would be visible. This fact is made obvious in some animation programs that do assume some level of default light when they boot (which really is a proper thing to do). The result is that a beginner using such a program will set up objects and go to render, only to find that the frames come out entirely black!

Other programs, such as Maya, offer default **modeling lights** that let you see what you are doing but that do not affect the rendering. These lights are analogous to the lights used by still photographers when they position their models on the set before they take a picture and trigger the actual flashlights that illuminate the film.

Still other programs include a certain amount of default light to start the user off, but strictly speaking this does not serve many professional animators, especially game developers, who prefer to start fresh with a dark set, just as a good cameraperson will enter a studio and only ask for house lights to avoid tripping on a piece of scenery.

As this book goes to press, a major divergence is occurring in the 3-D animation business. In the early days, 3-D lighting emulated real-world film-studio lighting. To light a scene, you loaded various lighting instruments—spot lights, fill lights, point lights—and attempted to make the result look realistic in the same way a film cameraman would on a film set. The results were pleasing, but no matter how good an animator got at imitating real life, the results were always somewhat plastic and lacking in realism.

The limitation of realism was well known to all animators and engineers all along. It was a matter of mathematics. In order to achieve realism, animators would have to abandon all the tricky lighting instruments and rely on the same lighting physics of the real world. In the real world, light enters a scene and bounces around a long time. It strikes a shiny surface and reflects. It hits a dull surface and scatters. It hits a transparent surface of a higher density and bends or **refracts**. Finally, it just gets plain tuckered out and fades into darkness. Animators always knew that if a computer were powerful enough to figure out all that bouncing around for every ray of light, the results would be astoundingly real. But in the early days of computer animation, such computer power was a distant dream.

Eventually, several developments coalesced into a small revolution in 3-D computer lighting. The first development was the advancement of the **central processing unit** (**CPU**) to faster and faster limits. The second advancement was the development of specialized visual rendering chips, both for immediately displaying an animator's tests on the monitor and for rendering the final result to high-resolution frames. The third development was the growth in the capabilities of video display cards fast enough to provide fast, real-time displays of the lighting and texture of a scene without rendering. This was accomplished by employing **Direct X 9.0** and **high-level shading languages** like nVIDIA's **Cg** and ATI's **Rendermonkey**. The fourth development was the authoring of more accurate and efficient rendering software that could employ the CPU and display circuitry to calculate the vast amount of data needed to calculate realistic lighting.

Today animators are beginning to fulfill their decades-old dream of photorealistic creativity, but we are still sitting at the fork of a technological road where both directions can still be chosen. If they prefer, animators can tinker around with the simpler, quicker traditional lighting tools, especially if they are in a rush or the results of the production do not require photorealism. They may prefer to employ more complex forms of lighting based on computer calculations of real-world physics.

Good, Fast, or Cheap—Pick Any Two

This is a well-known engineering maxim and it applies well to the status of 3-D animation lighting today. If you want fast and cheap results, go with the traditional lighting instruments (assuming you know them and are skilled in their use, which is worth learning). If you want good and fast, buy lots of the best rendering software and use the advanced lighting tools that employ real-world lighting techniques. If you can settle for good and cheap, use your computer's one CPU and the default rendering software that comes with your animation program. Employ the advanced lighting tools that LightWave, 3ds max, and Maya now include in their latest versions; set the scene to render before you leave for the weekend; and see what you get on Monday morning.

As of this writing, many third-party developers of rendering software, working closely with the 3-D animation program creators, are developing automatic lighting routines that do a fairly good job of executing all the lighting requirements as a function of the render stage, nearly eliminating the need to work with lighting devices within the program. Further advancements in rendering technology are beginning to place all the lighting solutions in the render software, so that eventually the use of traditional lighting instruments will become obsolete and lighting will become a simple, natural part of the render process and nothing more.

At its most basic and theoretical level, 3-D lighting, in emulation of real-world lighting, attempts to calculate the path of every ray of light in a scene from its origin to all collision results and finally to some end or fade-out point. Using such calculations, an animator need not attempt to imitate real-world lighting by imitating a Hollywood cameraperson, placing special-effect lights in a scene in order to create a realistic effect. The calculations can achieve this using only the lights that would, under practical circumstances, be in the scene. Such lights are called **practicals**, both in Hollywood and in the animation studio. Obviously, such calculations take a great deal of hardware assets and time. Over the years, developers have created methods that emulate this goal.

Radiosity

One of these methods is known as **radiosity**. Using radiosity, the artist merely lights the scene with the same lights one would find in reality (such as sunlight for an exterior, and a ceiling light and a desk lamp for an interior). The computer, employing radiosity, produces the lighting effects using a kind of shorthand. The shorthand employed by radiosity uses **patches**, or tiles of light, to compute a light transfer from a light source to the environment. Instead of calculating every ray of light, the computer employs a kind of averaging technique, dividing the light sources into tiles based on the levels of energy they emit while also dividing the surfaces into tiles based on the levels of energy they absorb. Then the program distributes that energy proportionately around the scene between the emitting light tiles and the receiving object tiles. Radiosity produces realistic results in less time than actually calculating the rays of light.

A more complex process than radiosity is **ray tracing**. Ray tracing, as you can imagine, is mathematically complex. Not only does the computer have to geometrically trace each ray's path, but the results of collisions with objects must also be calculated. Sometimes those objects will reflect the ray, sometimes the ray will be absorbed, and other times the ray will refract or bend like light in a glass of water.

Ray tracing often includes both forward and backward calculations of light in order to simulate all possible light paths and light interaction from light sources to the camera. This includes indirect lighting, light focusing, translucency, glossy reflectors, and even radiosity, without the need to manually prepare and combine each effect separately.

On its path, the light may change color, intensity, and position. Finally, all these calculations have to be compiled into a mapped grid of pixels such that the final frame of animation produces the result you anticipated. It is interesting to note that in most cases lighting calculations for such techniques as ray tracing, for instance, need only be calculated once for any given scene. These calculations are then stored in a kind of numeric map. Once the map for your scene has been calculated, the computer can proceed to apply that map to every subsequent frame in the scene. If you move objects, lights, or the attributes of either during a scene, however, the computer has to recalculate the map for every changed frame.

Monte Carlo Accounting

Most animators and the producers who hire them often cannot afford the time required for pure ray tracing; it would force them into bankruptcy, so the animators take a bit of a risk on getting the correct solution to mapping scenes accurately. This gamble is called Monte Carlo accounting. In the same way that a gambler may visit the principality of Monaco on the Mediterranean coast and throw a series of random dice rolls, so the animator throws a series of light beams into a scene. The gambler hopes his dice will produce the right number of combinations. The animator hopes his rays will hit something meaningful. Because fewer rays are applied in Monte Carlo accounting than would actually be employed in a precise calculation of the scene, the Monte Carlo approach produces results faster than pure ray tracing. Because actual light paths are being calculated, the result is often much better than radiosity solutions.

Photon Mapping

Photon mapping is a technique that takes two passes to calculate the lighting solution for each scene. The first pass employs a mathematical emulation of photons, which in the real world are the theoretical representation of a light's mass. In the computer animation realm, by using the properties of particles and emitting them as representations of energy from each light source in the scene, we are able to emulate photons. The result of this emission is a photon map. A second pass of calculations for the scene's final lighting solution is then made to calculate the caustic results of photon collisions with objects. This pass produces a caustic photon map. Using the two maps, the computer produces a map of the scene that is highly realistic yet moderately time consuming.

Pure Ray Tracing

Currently, pure ray tracing remains a subset of Monte Carlo accounting, simply because of the immense rendering requirements of using more rays. Really, no pure ray tracing exists, because nature virtually uses an infinite amount of rays to produce what we perceive as day-to-day reality. Pure ray tracing therefore always represents an exponential reduction in the kind of ray density employed by the Almighty. Many programs enable you to assign the density of rays employed in this technique. Be aware of the time required to render. Your computer has not crashed (well, maybe it hasn't crashed). It's just crunching all those numbers!

As a subset of Monte Carlo accounting, ray tracing is employed to calculate the most obvious results of emulating the real world: reflections. The application of pure ray tracing may not be really noticeable when rendering objects with low specularity, such as sand or cloth, but put a shiny Harley Davidson in the scene, and you notice really fast that the chrome doesn't look real unless some ray tracing is going down.

Therefore, program designers who offer ray tracing usually allow the operator to apply this render-intensive technique only where it matters most. Perhaps one day when computers get really fast, we'll see the emergence of pure ray tracing with high densities of rays becoming a reality.

Radiosity, photon mapping, Monte Carlo accounting, and pure ray tracing are some of the techniques employed in **global illumination**. Using global illumination, the animator accepts a different set of rules than those provided for in traditional or standard lighting. If a scene takes place in an exterior setting, the only global illumination instrument needed would be the sun (perhaps assisted by a reflecting surface to bounce light into a shadowed area). If a scene takes place in an interior setting, the only light really required would be the practical lights in such an interior: lightbulbs in lamps, candles, sunlight through the windows, and so on.

Now I hear you asking, "What about skylights?" Some programs offer **skylight** or **sky dome** lighting options that provide various levels of realism based on global illumination. In these options and in many global illumination solutions, the user may be offered the ability to enter the calendar date, time of day, and geographical location, whereby the computer will automatically calculate the precise position of the sun during the course of your animation.

In other words, as an animator entering the field in this era, you have the best of both worlds, considering the techniques and tools of advanced, realistic lighting are still in their infancy and have a long way to go.

Render Engines

As you know by now, a 3-D animation program consists of both the design function and the rendering engine. You design the objects and animations; the rendering engine makes the frames. Obviously, rendering is a highly mathematic function.

Early in the development of 3-D animation software, rendering was comparatively simple and merely translated all the user's specific instructions to the composed frames. As each software developer competed with others in the field, continual enhancements were added to each brand. Many of these enhancements were made possible by relegating larger and larger tasks to the render engine.

As noted in Chapter 5, "Objects and Surfaces," vast improvements in surface textures have been made possible by advances in the rendering components of the programs. Realistic fur and hair, for instance, are achievable due to mathematical algorithms that interpret the user's general commands ("I want long, shaggy fur here") into specific formulae that meticulously create each strand with the right length, texture, and weight. No longer does the operator have to create an individual two-point polygon replicated thousands of times and apply similar attributes to each one to create a field of grass. The render engine can do that automatically by choosing an appropriate **shader** for grass.

Similarly, the complexity of lighting is currently migrating from the user's skilled hands to the rendering engine's domain. Some traditional animators are resisting this trend, preferring to remain in direct control by using specific lighting instruments, especially when a unique, unusual effect is required. Many game developers, for instance, are loath to abandon their Hollywood-inspired lighting tools, but the advance of technology, its ever-improving capabilities, and the pressure of producing good work with a limited budget are driving everyone to accept the rendering engine as an indispensable tool in lighting.

Although each of the three rendering programs in this book has its own render engines, many animators migrate to plug-in renderers or completely independent rendering software. One reason is because each renderer has its own idiosyncrasies, features, and benefits. Rendering is a highly mathematical process that pits accuracy against speed. In order to speed up the rendering process, some manufacturers truncate the mathematical processes at a certain point, rounding off results and thereby reducing the time for calculations to take place. The compromise among speed, accuracy, and the necessity to invent better math processes is what makes one renderer behave better than another or produce results that appeal to animators in different ways.

In order to provide an adequate discussion on lighting, the available rendering engines must be covered, as well as the available traditional lighting tools. As a beginner, you must know how lights are used on a 3-D stage, because a scene will always need to be tweaked, and because the real world and the animation world often coexist in a film. Maybe you'll one day animate a *Monsters, Inc.*, where the real world is only glimpsed occasionally through a wooden door. When the director says, "Can we have a goboed key light on the wormy thing over there," you're going to have to know what a gobo and a key light are, and how to get them into your shot.

NewTek's Screamer Net

NewTek was one of the first companies to address the necessity of rendering animation over a network of different computers. Originally, they introduced ScreamerNet as a separate software product dedicated to rendering LightWave files, but they quickly realized the marketing advantage of including the render software as part of the whole program package. Since then, NewTek has continually developed ScreamerNet along with all the other tools in LightWave.

Although LightWave offers some of the most meticulous control mechanisms for their virtual lighting instruments, ScreamerNet has incorporated some superb, automatic lighting tools within the global illumination capabilities of the render engine. These tools enable users to experience the best of both worlds, individual lighting control and automated global illumination, all within one software package. Users will find all the advanced lighting tools offered by many third-party render software manufacturers right in LightWave's lighting tab, including global illumination, ambient color, ambient intensity, lens flares, volumetric lights, radiosity, and caustics.

Discreet and Alias | Wavefront

Discreet and Alias | Wavefront have historically worked closely as **original equipment manufacturers** (**OEMs**) with third-party developers to create plug-in rendering engines that are linked to their animation programs. These plug-ins offer rendering software that not only calculates the final frames, pixel by pixel, but that also enhances the final appearance of those frames by incorporating a wide range of effects.

One of the most recent developments is **subsurface scattering**, which is often used in medical images. It enables light to penetrate below the surface of human skin (and other similar surfaces) and then reemerge as a subtle kind of internal illumination. The effect produces amazingly realistic results and can even address such unusual skin parameters as oiliness, dryness, and pore size. Obviously, this feature is going to get a lot of notice from the makers of cosmetic commercials.

Rendering plug-ins with lighting tools are currently provided by Mental Ray (3ds max, Maya, and Softimage), VRay (for 3ds max), Cebas Final Render, and Brazil.

Mental Ray   Mental Ray is a product of **mental images** (**www. mentalimages.com**), an international leader in rendering and 3-D modeling technology for the entertainment, computer-aided design, scientific, and architectural industries. Based in Berlin, Germany, with a subsidiary in San Francisco, mental image's multinational staff includes many PhDs in such fields as mathematics, physics, and computer science.

Among its useful rendering capabilities, Mental Ray's global illumination permits a physically correct light simulation of **caustics** (see Figure 6-1). Caustics are light patterns caused by a refraction or reflection of light, such as bright spots caused by lenses or bottles, or underwater light patterns. **Volume caustics** and **multiple volume scattering** are also supported. Volume caustics are the effects of focused light made visible by fog or another diffusive object. Multiple volume scattering is the effect of diffused light, such as sunlight under water, light through clouds, or halos.

VRay   VRay (www.vrayrender.com) has been developed by the Chaos Group to provide a cost-competitive ($299–$799), quality-enhanced plug-in for 3ds max. VRay claims to be one of the fastest renderers in the field,

Figure 6-1
Example of Mental
Ray rendering
using caustics and
reaction-diffusion
textures.
Copyright mental
images, 1998.

exploiting all the assets of the host computer. It supports full, multi-threaded ray tracing in order to maximize the power of computers installed with multiple CPUs.

VRay accurately automates many of the traditional lighting and staging functions, such as the simulation of focusing (depth of field). VRay's caustics calculations allow property adjustments to be made on each object, changing the traditional approach from adjusting the lighting tool to adjusting the end result of that tool.

finalRender finalRender Stage-1 (www.finalrender.com) is, as of this writing, the only discreet, certified render software for 3ds max. It offers several unique features, such as an illumination model based on individual light particles measured accurately in 3-D space. finalRender Stage-1 is a true "external" renderer that replaces the 3ds max core renderer completely. No special work routines need to be added when finalRender is employed in 3d max. The user merely asks 3ds max to render, and final-Render takes over automatically.

finalRender supports global illumination, caustic light rendering, blurry reflections, photorealistic area lights, volume light rendering, and geometry-based direct and indirect lighting sources. finalRender also supports diffuse light interreflection, which renders a photorealistic image with proper light distribution.

Brazil r/s   Brazil Render System (Brazil r/s) is a 3ds max plug-in renderer from SplutterFish (www.splutterfish.com) featuring global illumination, image-based lighting, and full support of 3ds max's point lights and caustics. Brazil r/s brings its own specific point lights, area lights, and shadow algorithms to 3ds max users, as well as a photon mapping and **high dynamic range images** (**HDRI**) rendering.

Brazil r/s handles existing 3ds max scene files natively, requiring no special routines once a file is sent to render. It is also fully compatible with out-of-the-box and third-party 3ds max plug-ins. As of this writing, the engineers at SplutterFish promise to offer free upgrades and bug fixes to all registered owners of Brazil r/s, and owners may participate in free beta testing of new features as they become available.

Traditional Lighting Tools

If it can be said that the kind of light produced by a renderer is based on the objects being lit or causing light, and that traditional 3-D lighting is based on lighting instruments, we can easily understand traditional lighting as being composed of just three kinds of lights.

Because no standardization has been established in the terminology used by the various animation programs, and because each manufacturer is advancing different technologies, it is a bit difficult to categorize and name them here. Generally, the three types of light are as follows:

- **Ambient light**   3ds max calls this **daylight**.
- **Volume-defined**[1] **light**   3ds max calls this **photometric**. Maya's **volume** and **area** lights fall into this category, and LightWave offers **luminosity**.
- **Instrument light**   3ds max calls this **standard** light.

---

[1]In order to unify some of the terminology, I have chosen this term to define lights that are defined by a space in which the light is seen. This term should not be confused with volumetric light, which describes a light-scattering effect explained later.

Hopefully, as lighting, rendering, and the natural competitive interaction of manufacturers progresses, these terms and their underlying features will eventually coalesce and make the learning and understanding of computer-based lighting a bit easier.

You will notice as you read further that ambient and volume light are not, strictly speaking, derived from instruments. They predate (and perhaps predicted) the use of rendering software as a lighting control mechanism. Does that confuse you? Well, don't worry about that. It'll all be clear in a minute.

All lights allow color and intensity to be controlled. Most programs enable lights to be associated with specific objects, so that the light moves with or targets the object. The attributes of lights can also be keyframed so that the passage of time within the timeline can cause lights to change at specific points in the animation.

Ambient Lights

Ambient light usually comes from all directions and hits everything equally, although some animation programs enable ambient light to have direction. Generally, ambient light casts no shadows, because any normally shadowed area would receive as much light as the lit area, but some programs enable objects under ambient light to cast shadows if ray tracing is employed.

Ambient light is "cheap" light. Amateurs love ambient light because it does not ask for any specific target, attributes, or movement. It's just there, like sunlight, but is even simpler than sunlight, which, as you know, casts shadows and has direction.

When setting up an animation, placing stage pieces in the scene, moving objects around, and setting keyframes, ambient light can be a good tool in the same way house lights are a good tool. You use it before you get serious about lighting and then, when your players are onstage, you shut it off and start the hard work of making drama with light.

Drama with light is an odd concept. It sort of implies that light is a character, which it is in stage, film, and 3-D animation, but more so in animation than the other two. In animation you have enormous control over your lights, infinitely more control than in the real world, where lights are subject to the limitations of money, power, logistics, space, and convenience.

On a film set, you can't move the light in front of the camera. It would block the view. In a 3-D animation set, you can; the lighting instrument is invisible. On a film set, you can't have a billion watts of power. In *Star Trek*, the animators regularly created a supernovae and red giant suns with exponentially more power than any earth-bound instrument, by using the virtual power of CGI.

When lighting in 3-D animation, your only limitation is your imagination. Once you have created your scene and objects, lighting introduces a kind of iterative process that makes the workflow somewhat cyclical. You try some lights, render a test frame, examine the results, and think of some ways to improve those results. Then you change the lighting, render another test frame, and so on. Each rendering is an iteration, an attempt to improve the results of the last try or at least to learn what works and what doesn't. Once you have the lighting the way you want it, you move the scene to a different configuration and begin the iterative process all over again.

Maybe I was wrong when I said your only limitation was your imagination. You also have a limitation on time and, if the job is for pay, on money. How many iterations can you perform before the time and money run out? The better you know your tools, the more you can do.

Volume-Defined Lights

Volume light, area light, and luminosity apply to lights with the characteristics of objects or that are part of an object. Volume lights create light within a specific 3-D space defined by the shape of an object (usually a cube, sphere, cylinder, or cone). Within the space, the light can be graduated and directed. The light can come from one direction or be radiated from the outside or from within. An example of a volume light can be one that shines from a campfire or one that illuminates a room from sunlight through a window.

Volumetric light is an effect that can be employed to simulate visible light scattering and diffusion effects. Depending on your program, volumetric effects may be applied to volume-defined lights, including area and photometric lights, to simulate the real-world effect of light being diffused and scattered by such elements as dust, fog, smoke, and haze.

Photometric lights, offered in 3ds max, are another form of volume-defined lighting that accurately replicates real-world lighting instruments. Photometric lights use light energy values provided by the manufacturers of real-world lights to show and compute light distribution in a scene. Users can set the distribution, intensity, color, and other attributes of these accurately emulated real-world lights before populating an interior scene with them. (Discreet recommends that daylight lights be used for outdoor scenes and photometric lights be used indoors. They also specify that animators should not mix photometric lights with standard lights because the two types of lighting are not compatible.) Photometric lights are render intensive and rely on a scene being "light tight" to work properly. That means you can't have holes in your set, missing walls, or areas that are not completely designed, because photometric lights will cause the computer to use everything in the scene to create a realistic effect.

An **area light** is a lot like a **softlight** used by film and video camerapeople. An area light emanates from a rectangular surface, whereas the softlight is made of several bulbs in a square array covered by a rectangular diffusion surface such as white, spun fiberglass. You can move the area light around your virtual stage using the move-object controls of your software the same way a film gaffer moves a softlight around a film set.

Luminosity describes an object's internal glow. The term was likely coined by NewTek in an early version of LightWave and refers to an attribute now found in many programs. Using luminosity settings, an object in LightWave can be made to emit a glow that imitates subsurface scattering. It can be cranked up so that the object itself becomes a light source, such as a fluorescent bulb.

Some programs offer **linear lights** that, like NewTek's luminosity, act as light emitters with a specific shape. Linear lights usually function to emulate real-world lights, that employ an excited inert gas within a sealed glass container, such as fluorescent or neon light tubes.

Instrument Lights

Instrument lights closely match the lighting instruments used in the video and film industry. Instrument lights are broken down into **directional lights**, **spotlights**, and **point lights**.

Directional lights are intended to imitate light that comes from a great distance and is therefore not usually subject to decay or **falloff** (where the light grows gradually dimmer with distance). An example of a directional light might be sunlight. Directional lights are usually displayed in 3-D programs as some graphic representation of their real-world lighting function, such as a sphere with lines radiating from it in all directions.

Spotlights are, as the name implies, virtual representations of the film and video industry's lighting tool. Its graphic is usually self-explanatory: a cylindrical metal device with a glass front and barn-door baffles on each side. The graphic may also represent the shape of the emitted light, in the shape of a cone originating at the light and extending toward its subject.

Some programs enable the operator to view the scene through the light as if it were a camera. This assists the operator in accurately directing the light. Directional lights can also be parented or slaved to other objects, so that when the object moves, the light follows the object. Examples of a direction light would include a searchlight or light from the sun.

Point lights shine from a specific point out into all directions. Examples of a point light are a candle, a campfire, or an explosion.

Light Properties

Different types of lights have different properties, nearly all of which can be keyframed. It is important to note that changing a light's properties during an animation requires the computer to recalculate all the parameters of that light throughout the scene for every frame in which a change of properties occurs.

For example, let's say you are creating an animation in which the sun is rising. This can be accomplished by changing the color, position, and intensity of the sunlight during the animation. The computer will take a great deal more time to render the result than if you choose to have the sun at one position, one color, and one level of intensity. This fact of animation life applies to both standard and advanced software lighting tools. In fact, with advanced lighting tools, the time taken is exponentially greater than with standard or traditional lighting tools because of the higher degree of ray-traced calculations that must be executed.

As stated earlier, a light's color and intensity can be controlled. Color is usually selectable using a standard color wheel or a combination of color sliders, exactly like those used for object surfaces. Intensity is adjustable with sliders and numeric windows. Some programs use percentages to measure intensity, and some use real-world equivalents such as wattage; others use numbers that are relative only to other lights in the program.

The decay rate is a light property used by most spot, point, and area lights. As you know, real light tends to fade in relation to the distance from the light's source. You can assign a specific value that determines how fast a light decays in your scene. A graphic device that marks the limit of the light you are adjusting usually represents this limit in most programs.

Traditional lighting tools that represent spotlights usually offer a method of adjusting the **cone** or spread of the light from the lighting source. This cone is a direct imitation of the real-world effect created a spotlight's barn doors, four metal flaps on the sides of a real spotlight used by the lighting technician to shape the beam of light. The cones offered in some animation programs offer individual barn-door controls, whereas in other programs only the angle and distance of the cone are adjustable, the cone being symmetrical in all directions. Again, a graphic device, usually the outline of the cone, represents the border of the light.

Shadows

Odd as it may seem, one powerful property of lighting is shadows. One would think that shadows were a natural result of the lighting process, but in computer animation, shadows are determined within the attributes of the lighting instrument. Many animation programs set their lighting defaults to "no shadow," which is the simplest and least time-consuming setting. Shadows take time to render, regardless of how they are generated.

Several different types of shadows can be assigned to your lights depending on the program, the type of lights, and the rendering method. The processing of shadows usually implies the use of ray tracing, because an object interrupting the path of light causes shadows, but clever animation programs offer imitation shadow techniques that do not employ ray tracing and its consequent increase in rendering time. Maya, for instance, offers **depth map shadows,** which are calculated by using the relative positions of objects and surfaces in relation to the lights in the scene. The shadows are then added to the final render without employing ray tracing. Other programs call this same effect **shadow maps** or **projected shadows.**

Lighting Effects

Another form of shadow effect is a **gobo**. This term comes from the film industry and refers to a piece of metal or cardboard placed over a directional lighting source with holes cut in it to form a light pattern. Some gobos have horizontal slots to imitate light from a Venetian blind, whereas others have leaf-shaped holes to simulate forest lighting. Many animation programs offer gobos or **shadow masks**, which enable the artist to use a picture file (for instance, a JPEG drawing of black and white stripes) as a gobo.

Fog is a feature employed by most animation programs to simulate the effect of minute particles evenly dispersed within the 3-D space of your scene. As you know, fog appears in the real world whenever water droplets, dust, or smoke particles permeate the atmosphere. Although real fog can be created in an animation program by the introduction of virtual particles, this technique is extremely render intensive, and animators usually employ their programs' fog feature, which employs a standard set of mathematic simulations to create fog effects. These simulations are quite effective, allowing for density, grain, intensity, and the depth of fog. The result can be as noticeable as a searchlight piercing a cloudbank or as subtle as a gentle softening of the shadows cast onto a sun porch on a spring morning.

In real-life photography, **flares** are considered an undesirable side effect of lighting and photographic lens design. Cinematographers go to great lengths to avoid lens flares (the polygonal shapes of translucent color that repeat themselves across a scene when the angle of light is oblique enough to enter the lens). Animators, however, tend to employ lens flares and other kinds of flares as a means of building the credibility of reality. (I suppose if you have lens flares, it implies you used a lens!)

Flares come in a variety of patterns and densities, and in some programs, the tools to make flares may be scattered about in different tool sets. When creating lens flares, the software will usually allow you to determine how many lens element flares are stacked[2] and how large the effect will be in your scene. Other flares include circles and **star patterns,** and the program will enable control over the amount of points in the star, the density gradient of the circle, and even the presence of a **halo**. Halos are currently popular in large explosions, where a flare will be preceded by a horizontal halo, depicting the effect of the concussion wave of the explosion.

Glow is another lighting effect, often combined with flares. Glow can be used alone or be modified by a **texture map** or **random noise** modifier. When modified, glow and other lighting effects are distorted from their normally regular distribution patterns into patterns of varied intensity that are determined by the levels of black and white in the modifying source. In your chosen program, look for the capability to employ texture maps or random noise in lighting effects. Such tools will usually enable the selection of the modifying source graphic (or noise source) and the amount of the distorting effect. The distortion can also be varied in time by using

---

[2]Incidentally, the quality of a photographic lens can be determined from its flare by counting the number of stacked polygons. Each polygon denotes a separate lens element. The more elements, the more expensive the lens.

keyframes in order to add animated realism. Keep in mind that using keyframes in any lighting increases their render time.

Traditional Light Positioning

Real-world photographers employ a lighting technique known as **three-point lighting**. In order to properly illuminate, define, and model a 3-D object, the minimum amount of standard light would consist of three items: the key light, the fill light, and the back light.

The **key light** is the brightest of the three (set at approximately 50 percent of the total light on the subject) and is positioned either on the right or left of the object. This light, usually a spot, is focused tightly on the subject and is responsible for defining the most important details. With such a strong light cast from an angle, one might expect a great amount of shadowing. These shadows provide further definition of the object's shape, but they may prove to be too dramatic or may leave too much unilluminated to be left alone. A fill light is therefore required.

The **fill light** is the second brightest of the three (set at approximately 30 percent) and is positioned in front of the subject on the opposite side of the key light. The fill light is usually a soft light and is defocused broadly across the object. It is responsible for filling in some of the shadowing caused by the key light.

By combining the effects of key and fill lights, you can produce a wide range of effects that may enhance the viewer's understanding of the object being lit. When a subject requires drama, the fill can be reduced to enhance shadowing. When a subject requires beauty and romance, the fill can be increased in ratio to the key to create a smooth texture across the subject.

The **back light** is the third brightest of the three (set at approximately 20 percent) and is positioned above and directly behind the object. The backlight is usually a spotlight focused tightly on the subject and is responsible for defining the edge of the subject when viewed from the front. The back light serves to separate the object from the background, and therefore none of the light from the backlight should be allowed to illuminate objects behind the subject.

When lighting multiple subjects, some of the lights from one subject can be used to illuminate another subject if the angle and purpose of the lights allow for effective, multiple targets. In most cases, however, each subject should be treated separately with its own array of key, fill, and back lights.

Compositing

If you've been paying close attention and reading between the lines of this chapter, you may be getting the idea that 3-D lighting and rendering entail enormous compromises. Well, guess what? You're right, but don't despair. As you get acquainted with the tools of your program and the available plug-ins, you will no doubt stumble across lots of work-arounds and the techniques of experienced artists who have come before you.

One of the greatest work-around tricks in quickly developing scenes of great realism is to master the art of compositing. Compositing is done outside the 3-D animation program using—you guessed it—compositing software. Three excellent examples you might explore are **Adobe After Effects** (www.adobe.com), eyeon **Digital Fusion** (www.eyeonline.com), and discreet's **combustion** (www.discreet.com).

Compositing involves three basic steps. First, you use your 3-D animation program to render your complex scene in layers. Second, you tweak the layers. Finally, you recombine these layers into final frames using the compositing software.

For instance, you may have a character walking through a scene that requires a higher level of rendering than other scenes. Behind the character is a row of torches burning in sconces on the wall. (Does this scene sound familiar? It does to me.) A smart animator, such as Jack Ehrbar (see Chapter 2; page 59), would render the torch flames in one layer and the character with his ray-traced shadows in another. The rest of the scene could be rendered in still another layer. Without the render-intensive flames, character, and shadows, the scene renders pretty fast. The animator could then tweak the scene (without the render-intensive items) to perfection and then render a final sequence, calling it, perhaps, "Background."

Next, the animator could tackle the flames, rendering these in a separate layered sequence and tweaking them to perfection (or at least as far as the producer will allow). Then he or she could save that final sequence as "Flames."

Finally, the animator would set up the character to render all by itself with ray-traced shadows and then break for lunch. When he or she returns, the animator could tweak the character and shadows another time and then take another break. Eventually, however, the animator will settle for the best render he or she can get and save this final sequence as "Character."

Then, as the producer breathes humidly down the animator's neck, he or she will open a compositing program, import the three sequences, and combine them. Compositing programs offer a vast array of image manipulation tools that can assist the animator in further perfecting the final composition, even though each individual layer may still have problems that could be corrected with further tweaks and renders.

In fact, the producer might have a high colonic event and demand that the animator go back to rerender a layer, but most of the time the animation team can "fix it in the mix" by adjusting the color, density, and position in the compositing software. Because the compositing software is only using 2-D images that are limited to the final resolution of the animation, the iterative tweak-composite-render process in the compositing software is much faster than the iterative process of a 3-D animation program.

Therefore, when planning your final rendering workflow, always consider the time-saving aspects of compositing. By intelligently layering your scenes, you can often save as much as 50 percent of the total production time.

Rigging and Animating Characters

3-D animation consists of many tasks and the most challenging, complex, and vocationally fruitful task is character animation. **John Kundert-Gibbs** and **Peter Lee**, in their excellent series, ***Maya Savvy***,[1] describe the attributes of an animator, concluding, "A good way to discover if you are an animator is to go through a whole animation project and ask yourself which part you enjoyed spending time on the most."

If you've been reading this book sequentially (and I know a few sick individuals who actually do that) and working on the demo tutorials for each program, you will by now have acquired the skills to create objects and scenes. You will also have added attributes and lights, and perhaps animated simple sequences involving objects and scenic elements.

If, like me, you bounce around in a book, read the last chapter first, dogear pages for later concentration, sketch in the margins, email the author with related trivia, and otherwise carry on like an aimless beggar of information, you'll probably have come to the conclusion that character animation is composed of three major tasks: creation, rigging, and animation, with animation being the most fun of all.

Character Creation

I'm sure, whether you are the anal-retentive sequential reader or the dysfunctional, attention deficit disorder-addled nonlinear reader, you can easily figure out how to connect all the balls, boxes, cones, and cylinders into a character that suits you just fine. Each of the three demo versions of the animation programs takes you down this path in its own individual way.

It's just a matter of taking a primitive object such as a sphere and using the **control vectors** (**CVs**) to push and pull the sphere into something like a face. You do the same to a box to make the chest, a cylinder to make the arms and legs, smaller boxes for feet and hands, and smaller cylinders for fingers and toes. Using the various welding and merging tools of the animation programs, you take all the elements and merge them into one complex object.

Oh, you think there's more to it? Well, you're right. By merging primitives, you can easily create elemental cartoon characters, but highly complex characters such as the title character in *The Scorpion King* or Dr. Aki Ross in *Final Fantasy* require a bit more effort.

---

[1]*Maya 4.5 Savvy*, 2003, Almeda, CA: Sybex, p. 253.

Attribute Lists and Sketching

Generally speaking, to create a complex character with personality and detail, you should start out by sketching your character the old-fashioned way: on a large 18- by 24-inch pad of sketch paper using a soft lead pencil. Grabbing a sheet of printer paper and a felt-tipped pen might work if you're the world's most brilliant artist, but even the world's most brilliant artists avoid cramping their style by using a large surface and an easily erasable drawing instrument.

If you're working for a client who has given you a specific list of attributes, such as happy eyes, big feet, and Mickey Mouse gloves, you should start out with a list of these attributes at your side. Even animators who start out creating a character for themselves often write down the attributes they want before they begin to sketch. Animation can range from a solo enterprise to a team effort requiring 30 or more artists. Either way, a simple text list of attributes helps you focus on the character you want to create.

Once you start sketching, don't limit yourself to any particular view or angle. Just get it down on paper. Make several versions, ranging from simple to complex. If you are working for a client, it helps to make several simple choices and show them to the client for approval. This allows the client to choose his or her favorite before you invest too much time adding details.

Incidentally, although most sketch artists like to draw their characters in appropriately dramatic poses, the correct way to draw a character for eventual input into an animation is in the **initial pose** stance. For a person, this would be with the arms outstretched to the sides and the feet placed flat on the floor shoulder width apart. When you scan the image into your computer and model it in 3-D, none of the character's details will be hidden or overlapping.

A student once asked me how a butterfly character would be posed. I brought back a biological specimen spread out with pins on a cotton bed and described how such a character would have to be sliced in half, severing the wings. The wings would be scanned in an open position, whereas the body would be scanned with its legs spread open.

Another student asked about a beetle, which obviously has several layers of wings and a folding carapace. Hey, you're going to have to scan flat art into a computer and trace it in three separate views. Maybe you're going to have to take out a scalpel and dismember the bugger to illustrate each part effectively. Not a bad exercise when learning how to imagine something that doesn't exist in the real world. Novice animators too frequently choose to design characters that are easily depicted in three views. Thinking about beetles gets you thinking competitively.

Case Study

I recently had to supervise the creation of a cartoon lion character that would be featured in a series of used-car commercials. The client was extremely picky and our first 10 sketches went nowhere. We'd try one thing and the client would say the opposite. We'd develop the opposite and the client still wasn't happy. Eventually, we gave up drawing completely and went searching in Google for "lion" and limited the search to graphics. We then created over 100 lion sketches and sent them all to the client with the request that he eliminate any sketch he didn't like (the majority) and tell us what he liked about the ones he didn't throw out (see Figure 7-1).

This process tended to annoy the client a bit, probably because, working under a flat price budget (an average based on dozens of cartoon projects), he figured he'd have some fun at my expense. Now, asked to specifically define his requirements with a wide range of examples that cost me nothing to deliver, the client quickly settled on a list of specific attributes we could use to create a sketch he liked.

Figure 7-1
On of the first sketches created by artist Jack Ehrbar in the process of achieving client nirvana for a used-car commercial

Not to say that our job was done at this point, because his pickiness included the way we illustrated each specific attribute, but after a month or so, we had a character he liked (see Figure 7-2). But then he had to present it to *his* client! To this day, we still haven't gotten approval on the character, by the way, but that's another story, and it's a happy one because I got my first payment on the job, which more than compensated for all the hard work.

Once your general sketch meets the approval of your client or your own requirements, it's time to start some clinical aspects of illustration. Eventually, you're going to have to convert your sketches to 3-D computer elements. You will be expressing your character in front, top, and side views. Therefore, it's a good idea to start sketching your character from such directions.

Once I had an occasion to interview for some work at MTV and one of their supervising editors visited my shop. The project they had in mind was

Figure 7-2
The final lion character as accepted by the client. Six hours were spent just getting the scepter right.

a portion of the *Beavis and Butt-head Do America* movie. The producers thought that the acid trip that Beavis and Butt-head take during the movie might possibly be done in 3-D animation. Naturally, I asked for three-view drawings of the characters.

Turns out, I accidentally stumbled on one of the sore points about the two poor lads. In their creation, **Mike Judge** had not given a great deal of thought to the profiles of his characters or the in-between stages between full face and profile. If you examine the characters closely, you'll see this weakness, which never caused Judge much heartache or seemed to affect the popularity of the show. It's just that a basic design flaw was permitted to exist that does not lend Beavis and Butt-head to be converted into 3-D animation. So the idea of a 3-D acid trip never got very far.

You, of course, will not fall prey to the same design flaws as Mr. Judge (I hope), but you may be as successful or better (I also hope) when you design your character with a well-developed front, side, and (if you're particularly enthusiastic) top.

Do you remember that intern, Heath Vincent, I introduced you to in Chapter 2, "From Concept to Screen: The Workflow Process"? In that chapter, we left off with the **character lines** of the knight and the dragon. Now let's see how Heath fleshed out his character in developmental drawings.

The first step is to create a concept drawing that incorporates the elements of the character lines you outlined (see Figure 7-3). The next step is to create full-body drawings of the character from each viewpoint (see Figure 7-4).

Notice how the character is clothed (he'd look pretty scary naked!) and equipped with various tools of his trade. These items cannot be forgotten in the character preparation process. The wise animator sets aside time to illustrate all the necessary items that will be required to equip the character. Sometimes these are just for concepts and sometimes you will actually scan and model them (see Figure 7-5).

Still not content with having created a complete compendium of his character, Heath went one step further before turning on the computer. This involved drawing detailed examples of the character's expressions. For the dragon, this was not much to worry about, but the knight needed a full repertoire of grins and growls (see Figure 7-6).

Notice how Heath went far beyond the basic requirements of orthogonal expressions and depicted the weapons, clothing, and other details of his character. Visualizing these elements helps bring your ethereal dreams into reality. Drawing them disciplines your professional hand and prepares you to develop the character on the computer as well as sell the concept to a pro-

Figure 7-3
Figure 7-3
Character concept
drawings of a
knight and a
dragon from
which animator
Heath Vincent
created his full-
body drawings

ducer. It also gets you ready to build a complete character ensemble, which
may end up being a very profitable endeavor (see Figure 7-7).

Inputting

The process of inputting your drawing into the computer is somewhat different for each program, and in the interest of brevity I will describe the
process in general terms. You can then explore each of the CD program
demos and tutorials to gather specific experience. Before you actually get
started inputting your design, you should consider some basic preparations
that, when built into your standard workflow, will produce consistently professional results.

Setting up Project Directories

At first, creating a character may not seem like a complex affair, but in a
short time you will have created a vast amount of files and data types. Just

Figure 7-4
Four of several
drawings Heath
made of his
characters, prior
to inputting them
in the computer

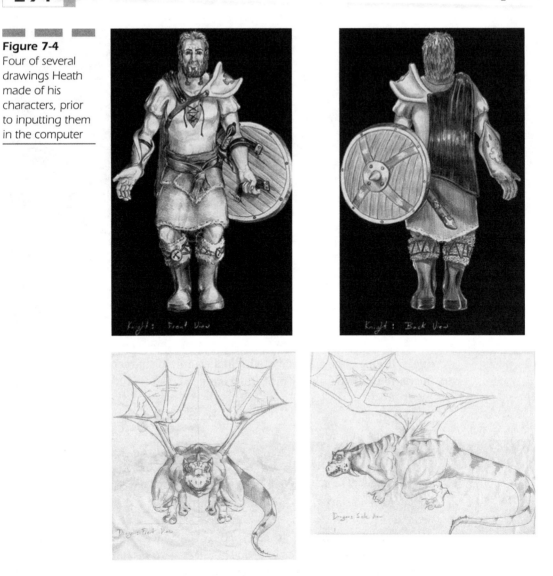

about the time you realize this, it will be too late to sit back, take a day off, and easily reorganize all the files. It will be too late because you will have already forgotten where you put most of the files, some of the names will make no sense, and, worse, some of the names may be duplicated. Right at the get-go, decide to create a logical tree of directories for storing all your work.

If you are working at someone else's facility, say, as an intern, start out by naming a directory under your name. That's the convention in my shop,

Figure 7-5
Samples of the knight's clothing and weapons

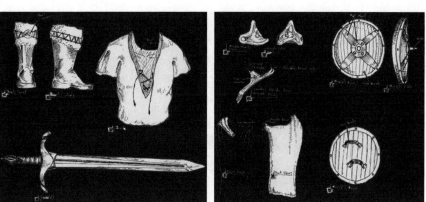

Figure 7-6
A few of the facial expressions Heath created for his knight character. You may require a lot more for lip synching.

Facial Expressions

Default   Disbelief   Pleased   Disappointed

Angry   Intrigued   Horrified   Bemused

where we have about three interns a year cycling through. Once they're gone, the understanding is that I can wipe their directory whenever I need the space.

Under your name directory, create a different subdirectory for each project. Within that project directory, create folders for models, scenes, meshes, maps, graphics, or whatever you need—as you need them. Then deposit within each directory only those items that fit the category name. When you get to naming character parts, use a common convention used by most 3-D animators. Choose a name for the character, such as "Joe." Then name the specific anatomy parts, such as "rleg" for the right leg, and provide names within the subanatomy, such as "rleg_toe2," so that the entire name of the second toe on the character's right leg would be "Joe_rleg_toe2." It is

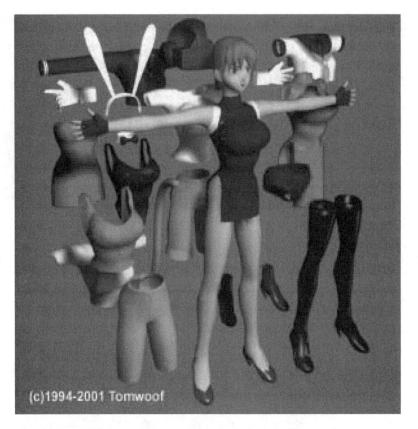

Figure 7-7
An example of a well-coordinated character ensemble designed by Japanese animator Pang Sie Piau, AKA Tomwoof, for display on Turbosquid (www.turbosquid. com), an online agency selling 3-D models

(c)1994-2001 Tomwoof

helpful to keep a notebook beside your workstation for listing all these names and conventions.

Eventually, you may want to perform automated procedures on a large group of files. By maintaining a strict set of naming rules, you won't find holes in your process later by virtue of having a section of your character with the wrong name. All this may sound dull and stupid now, but three weeks into the project when you need to find the texture map for the toenail, you'll be happy you did this.

Another difficult part of character creation is keeping track of where each character element comes from. I learned the hard way, using a **non-linear editing (NLE)** system, that files can be gleaned from a lot of places when used in a project. If other people use those files (and delete them when they're done with their projects), you can risk serious losses. Therefore, no matter where you obtain textures and generic elements, always make a copy and stick it in the right subdirectory of your own named project. That way, it'll always be where you need it.

The Massing Model

Prior to doing any specific modeling, you may want to consider the advantages of building a quick **massing model** (see Figure 7-8). A massing model is usually composed of simple components—boxes, spheres, and cylinders—and is placed in a scene to figure out the elemental basics of the animation you envision.

These basics might be related to a specific storyboard, allowing you to see how the character would fit in each scene and if the storyboard works in the 3-D world. Often, a storyboard artist, thinking in 2-D, will neglect certain aspects of 3-D realization that can only be revealed by seeing scenes built in 3-D. Using a quick massing model allows you to create a 3-D storyboard with little effort, thereby eliminating some problems before you invest too much work.

Figure 7-8
A simple, human mass model made from polygon primitives in Maya that can be used for poses in a number of storyboard situations to test angles, positions, and trajectories

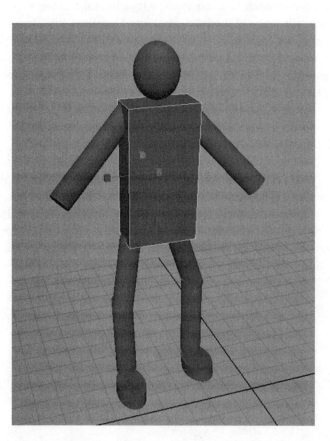

Other facets that can be examined using a massing model include the proportions of your character in relation to the scene, the best poses for the character in each scene, and basic motions for the character in relation to the soundtrack, music, and other time-related limitations. Stated another way, the massing model assists you in creating a 3-D version of your storyboard: minimal work, maximum results.

The first step is to scan the drawings into your computer using a common scanning device such as an HP Scanjet 5500c, for instance (see Figure 7-9). Nothing fancy is required. Simply set the scan pixel ratio as required by your animation software. This will usually be around 720 × 480 for NTSC, 1924 × 1080 for *high definition* (HD), and so on.

Scan at least the front and side views. If you've done a top view, this will also be helpful. Bring the respective views into your animation program as reference graphics. At this point you must decide how you are going to build your character.

You will remember (if you've been reading this book's chapters in numerical order) back in Chapter 5, "Objects and Surfaces," that 3-D animation uses four types of modeling: polygonal, spline, subdivision surfaces (patch), and *Level of Detail* (LOD, also known as parametric modeling). As you know, each method has its advantages and disadvantages for making various kinds of objects, but here is a short review in terms of their application to characters.

Polygon Characters

Polygons are quick but historically have yielded angular, nonorganic results. Everything in a polygon character is made up of small, flat surfaces. If you need a smooth surface and the capability to closely examine a character's details, polygons may eventually reveal their coarse edges as undesirable defects in your character's surfaces. Polygons also cost you render

Figure 7-9

Even a high-performance scanner like the HP Scanjet 5500c costs under $300.

time and/or creation time. You have to make them small enough to be invisible and then wait for the computer to deal with all the complex data. Recent developments in technology, such as subdivision surfaces, NewTek's **MetaNURBS,** and discreet's **MeshSmooth,** have breathed new life into polygon modeling, but pure polygons are still useful for designing characters that are basically composed of polygon primitives, such as robots and aliens based on crystalline architecture.

The computer gaming industry still uses polygon modeling in order to reduce the real-time computational requirements of characters in TV set-top hardware. Gaming is testing its limits, however. Even as we speak, one of the largest game developers is experimenting with new ways of applying lighting without addressing the individual surfaces of the polygon. (I'll leave it up to you rec-room geniuses to figure out how this is being done.)

Spline Characters

Spline modeling produces great results, especially for creating realistic, organic characters that are easily updated. Because splines are mathematically produced from unseen influences, they are fundamentally resolution independent. This means you can zoom in on a spline and not see the individual planes or surfaces. Using smooth, natural curves, spline models are better than polygon models for creating organic characters.

Remember, several types of character-building programs are based on splines, and most 3-D animation programs will offer more than one. These include B-splines, Bezier curves, and *Nonuniform Rational B-splines* (NURBS). The drawback of splines is that they are not easily cut apart and rewelded. This makes them difficult to use as elements in large, complex characters.

Subdivision Surfaces Characters

Subdivision surface (also known as a patch or editable mesh) modeling requires more memory and render time than splines or polygons. They are also more reliable when stretched and bent, and are better suited to stitching together large, complex characters.

The surface of a patch model usually consists of a matrix of either polygons or splines (depending on the program), which, unlike splines, makes them easy to cut into the elemental shapes that are needed to build characters. Once the character is built, moving the control points that define the shape of the patch can easily reshape it. Unfortunately, patch models bring back the old drawback of polygon models in that the basic element is a planar polygon or a spline of limited resolution, which is revealed to the camera if zoomed in too closely. Patch modelers therefore find themselves destined to create several versions of their characters to suit various focal distances.

Level of Detail (LOD or Parametric) Characters

LOD or parametric modeling usually employs spline-based mathematics to enable a character to be resolution independent and represents, at this time, the latest development in modeling. This advancement is made by allowing the spline and its related surface elements to be continually redefined as a function of the distance from the character to the camera. In essence, LOD employs the camera as an integral influence (along with the control points) on the object's mathematical structure.

LOD characters are easy to build and manipulate, because they employ all the best attributes of the earlier modeling technologies. Because they are resolution independent, they are easily examined with your program's viewport's zoom commands, and they respond well to deformers that make a character appear to move its limbs.

The only disadvantage to an LOD character is that it will most likely only be able to reside within the program that created it (at least for now). If you plan to later export your character to another program or, worse, to a set-top box gaming environment, it will most likely have to be converted to another modeling form. Consult with your program supplier before investing time building an LOD character if this is your end goal.

Modeling the Character

Now the choice is yours. Converting your scanned drawings into the computer is a matter of matching your 3-D model to the drawing you scanned.

Polygon Modeling

If you are working in polygons, start by choosing primitives that best match each element of the sketch. Most 3-D programs enable you to modify the surface of these primitives by using tools such as Stretch, Scale, Twist, and Move. Finer tools are also available that enable you to choose individual surface points, groups of points, or entire areas and arrange them to match your sketches. By merging the various primitive shapes, you end up with a unified character that does not have any surface holes, problematic polygons, or duplicated surfaces or points. Now you're ready to "rig" the character.

Spline Modeling

Using splines, B-splines, and NURBS, the process is more precise than polygons. Spline models consist of subobjects that include curves and surfaces. The face of the character, where most of the dramatic expressions will occur, can be carefully traced from your sketch using contour curves. Some programs (such as Maya) enable the user to create CV curves by playing individual points in the view. Others require that a curve be created and then "massaged" into the required contour by moving the individual VCs.

The beauty of spline modeling is that very little data is required to define a surface. As you will remember from Chapter 5, a CV curve can be composed of as few as three points. By adjusting the **weights** or "bending power" of the CVs, you exercise more influence on the curve, thereby achieving results that in other forms of modeling might require many points to smoothly define. Consequently, using a minimum of three curves and the appropriate weights, you can define a rather complex but smooth surface.

This is no free lunch, however. Larger weights tend to add to the computational complexity of the object, although not as much as an equivalent amount of points in a polygon or patch model. Spline modeling nevertheless is usually the most data-efficient system. When a spline character is animated, however, the efficiency of a NURBS object declines. Animation requires distortion of the object, and distorting NURBS requires changing the CV weights, thereby increasing computation.

Begin building a spline character in one view, creating enough curves to define the character, but keep in mind that every CV adds computational time. Start off by spacing your curves out in one view, such as the side view. Then switch to the perpendicular view, the front (where the curves will look flat), and modify the curves to fit the contour of that view. Constantly check your results in the perspective view. When you have a sufficient amount of curves in place, loft a surface.

A common misconception about lofting a surface on CV curves is that the artist must create a network of intersecting curves to define the polygons, which in turn define the surface. This is not so. Surfaces can be lofted quite well between two parallel curves that never touch.

In most cases, you would create only half the object, mirror it, and weld the two halves together to form a whole character. Conscientious artists would then add some asymmetrical differences on either side of the character to mimic natural appearances, which in complex creatures are never perfectly symmetrical. Editing the curves and manipulating the control vertices may accomplish this.

Once the basic structure of the character is complete, you can add additional surfaces by adding additional CV curves. These curves may become part of the original structure or they may be individual, dependent surfaces with their own attributes. Such dependent surfaces are useful for creating eyes, eyelids, lips, or ears that may later be controlled to create character expressions.

Of course, a tool like **Immersion's Microscribe** can assist you in capturing CV points in 3-D space by using a plastic or clay model, thereby eliminating the need to use 2-D sketches and views.

Patch Modeling

Patches are polygonal shapes created by the intersection of three or more spline curves and the application of a surface across that polygon. Because each intersection is a CV, patches can be distorted as needed to create movement or character expressions.

Because patch modeling using subdivision surfaces is a combination of polygon and spline modeling, animation programs tend to employ two stages of patch modeling. The polygon stage of using primitives to structure the coarse aspects of the character allows for a quick assembly of the body, limbs, and digits, whereas the spline stage allows for a direct tracing of facial details and other areas of the character that might need refined expression. The modeling routines in LightWave's **MetaNURBS** are a good example of the polygon stage, whereas the **spline patching** tools represent the spline-modeling stage.

The ability to cut and paste sections of a character shaped by splines but defined by polygons makes patch modeling one of the most satisfying modeling methods in terms of getting the details you want in as short a time as possible. Many animators find patch modeling the most satisfying way of creating a character because the results are smooth and clean, yielding high degrees of detail where needed, with the capability to apply evenly distributed and controllable texture maps.

LOD and Parametric Modeling

Parametric modeling is an advancement of an important characteristic of subdivision surfaces and patch modeling. As you know, the level of detail in a character is based on the density of data that composes that character. Just as the density of pixels determines the resolution of a 2-D photo, the density of polygons, spline curves, and CVs determines the resolution of a character's surface and detail. Obviously, the closer you get to a character, the more resolution becomes an issue. A character designed to look great at 5 meters from the camera may show significant "jaggies" and planar artifacts at 5 inches from the camera.

Before parametric modeling, animators who wanted to incorporate wide shots and close-ups of a character had to design separate, replaceable characters for each situation, modeling each to accommodate the needs of the shot. Today parametric modeling allows the resolution of a character to vary according to the needs of the animator by changing the level of detail (hence the term LOD). This is usually achieved by subdividing the patches (and employing sophisticated averaging algorithms) as a direct relationship to the distance of the character from the camera. If the camera is far from the character, the surfaces are not subdivided. As the camera draws near, the LOD subdivides the surface into even smaller units, thereby maintaining smoothness and detail.

As of this writing, LOD or parametric modeling is offered primarily as a third-party plug-in, with the exception of Maya, which includes LOD modeling. **Softimage XSI 3.0** incorporates the capability to create reference models and then set them for high, middle, and low resolutions. LOD modeling represents one of the most important, cutting-edge aspects of 3-D technology at this time. Considering the usual level of competition between animation programs, it will not be long before LOD modeling is offered in all the leading programs and becomes a ubiquitous feature within the patch-modeling workflow.

Useful LOD applications are Right Hemisphere's **Deep Exploration** and Viewpoint's **Interchange 5.0**. As of this writing, LightWave users can download a free LOD plug-in at http://amber.rc.arizona.edu/lw/lod.html. 3ds max (and LightWave) users can also find LOD plug-ins at www. darksim.com/html/body_dt2_plugins.html.

LOD is also a competitive feature in video cards. **Nvidia**, for instance, has produced several graphic display chips that will both display LOD to the monitor and assist LOD rendering (see Figure 7-10). Check out some results on the Nvidia **GeForceFX GPU** at www.nvidia.com.

If you can't find the appropriate tools, LOD or otherwise, in your chosen animation program, you can easily translate your models from one program into another. For instance, if you created animation data and meshes in LightWave and wanted to export them into Maya, you could use **Okino's Polytrans**, which is capable of transferring a wide range of models and their associated data between 3-D animation applications (see Figure 7-11). Polytrans can be acquired as a plug-in to LightWave, 3ds max, and Maya, as well as a number of other animation programs. It can also be obtained as an independent program.

Rigging

I remember one evening years ago when my daughter, long before she became a hard-body West Pointer, suddenly broke down, bawling her eyes out as she held her favorite doll. My wife and I ran to her room, alarmed, and asked what was wrong. "Mommy," she wailed, "I just realized that all my dolls were made in factories."

Figure 7-11
The working interface of Okino's Polytrans program running as a plug-in in Maya. Note the wide range of import and export programs offered in the right column.

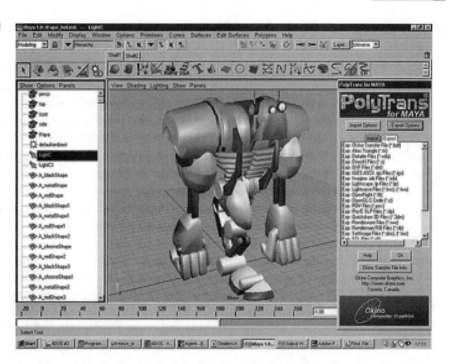

My wife, stifling an explosive belly laugh, blurted, "What did you think? That I gave birth to each of them?"

Also suffering the pain of mirth containment yet remembering, strangely, how I had suffered the same realization in a similar way (not with dolls, but with my favorite TV character, Willy the Worm), I managed to shed a sympathetic tear and gathered my child in my arms. "Here, sweetie pie," I said, "just take your Ñaña and hold her like this," and I showed her how, with two hands, she could control both the arms and feet of the doll. "Now you can make her walk and move her hands like a real person." My thought was to bring her through the crisis the same way I came through it: to realize that more magic exists in making magic than in watching it. And what do you think happened?

"Bwaaaah!" she wailed. "Ñaña comes from a factory!"

Clearly, some people are destined to become courageous defenders of our nation's democratic way of life, while others of us are destined to rig and animate computer-generated characters. If you're of the latter type, here's what rigging is all about.

The colloquial term, **rigging**, refers to what must be done to an object in order to prepare it to become a character. You understand that a character is just a big object, right?

Give Me Some Skin

You might think that the process of stretching skin over a polygon, spline, or patch character would be a simple affair. For the most part, it is, but animators are a discontent bunch. As soon as something gets simple, they make demands that get the whole thing complicated again.

In order to make a character move realistically, animators have to distort the shape of the character. Now you can go out and find some generic characters that are already skinned and rigged, and you can actually perform some tutorials or demos where you pull a certain screen controller and the character starts moving. That's great. You can even start to keyframe poses in the timeline and see the result of your animation. Well, that's exactly the way it's done.

But all this realistic action doesn't happen automatically. If you ever want to create your own characters that are capable of realistic motion (or if you prefer surrealistic motion), you are going to have to learn how to skin and rig your own creations.

Because character animation is a combination of distortion and proper skin control, the act of skinning a character involves tying the skin elements to the controllers of the distortion. Therefore, skinning and rigging (also known as **adding bones** or **building the skeleton**) go together as a two-step, cyclical task.

Skinning is most critical around the joints of your character, where the skin folds or expands to create the illusion of flesh, bone, and muscle beneath the skin. Times were, nobody minded much if, when you bent the arm of a character, the inside surface of the joint flattened like a straw. Hey, it's an animation, we'd say. Then some clever dude decided to create tools to control exactly how the skin is affected by movement. These tools provided a linkage between the controllers of the skin (the vertices) and the controllers of the movement (the bones or skeleton).

This linkage is handled differently in each animation program, but the generic term used by most programs is **weighting the skin**. 3ds max, for instance, provides a **gizmo** that paints weights onto the vertices (see Figure 7-12). Each type of weighting provides a means by which the artist can determine how much of an influence a bone has over a particular vertex or zone of vertices. For instance, on the inside seam of an arm, you would want a high degree of influence, so that when the arm bent, the skin would properly deform in a tight crease. At the elbow, you might want a slight influence to cause a bit of stretching and twisting of the skin. However, along the outside of the thigh, you might want very little or no influence as the knee bends.

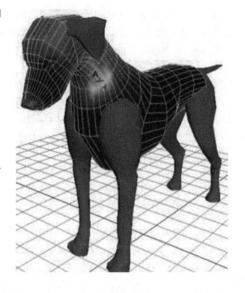

Figure 7-12
An example of painting weights on the skin of a dog character using the Paint Skin Weights tool in Maya. Photo courtesy of Alias|Wavefront.

Other skin-weighting features offered in many programs include **bulging, flex,** and **bounce.** These are controls that provide **soft body modifications** to a character or object. Imagine, for a moment, a really fat character who is startled. He spins around and freezes. What happens to his belly? Right! It continues to spin around after the skeleton has stopped! It goes so far and then stops and reverses! Maybe it goes back and forth a dozen or so times in ever-decreasing oscillations. This effect is good for a laugh, especially if the character notices it and grabs his belly in an attempt to stop the sway so he can concentrate on finding out what caused the startle.

This difference between the momentum of the skin and the position of the bones is a controllable relationship between the two entities. Elaborate mechanisms are offered to emulate **mass**, **inertia**, **oscillation frequency**, and **limits of oscillation**. This effect can also be applied to objects to impart a realistic or comically exaggerated flexibility, such as an arrow hitting wood at an oblique angle so the feathered end vibrates. Would you impart a tight, realistic vibration or a comical wobble, accompanied by a sound effect?

Budgeting Details

When you plan the design of your character, it is useful to carefully consider each scene in which the character appears. Many novice animators start out by designing a character without any idea of how that character will be used in a dramatic presentation. Essentially, such animators are creating a generic character, which is not a bad thing, but it may lead to a lot of wasted work.

A generic character can be used to demonstrate your abilities as a character designer, or you may want to specialize in this field and build a gallery of characters you might offer for sale (or rent) with the idea that the purchaser would order specific enhancements and detailing. In such cases, it is wise to rig such characters in a minimal fashion. You might even want to use generic bone structures as supplied with your software and adapt your character to these generic structures. In such a case, most of the rigging work has already been done for you, often for free.

On the other hand, if you are designing a character with elaborate boning and weighting, such as a Hindu goddess with four arms and individually boned hands and fingers, you may want to take the time to examine which scenes may actually require details and which may be best served with coarser controls. For instance, a long shot featuring a Hindu goddess posed on a throne with very little movement may require minimal rigging, whereas a close-up of two of her hands wrapped around the handle of a Louisville Slugger will certainly require some meticulous fiddling.

By examining the requirements of your storyboard, you may find you will need several versions of your character, some more detailed than others, and some featuring only specific body parts. In the same way you learned to model your scenic elements and objects with details appropriate to their roles and physical placement in a scene, you must also learn that an efficient analysis of a character's requirements may yield profitable savings in time. Time saved by eliminating unnecessary details is time that can be spent improving the areas of your character that will be seen, or maybe by taking your kids to an art museum.

Bones of Contention

Once you have a character designed to your satisfaction, it's time to get it out of that uncomfortable crucifixion pose and into some action. In order to do this, you will have to instruct the computer to deform the character. Novice animators are often surprised to hear that characters don't actually move around automatically like puppets. That could, of course, be arranged, but the results would be puppet-like and not very realistic.

As you know, a puppet is composed of fixed limbs and hinged joints. You move it with strings by pulling on the limbs and rotating them in one or two axes. Our 3-D characters are much more complex than simple puppets, but the complexity comes from just three advancements. The joints can rotate on three axes, the limbs can deform (also in three axes) in synchronization with the movement, and a virtually unlimited amount of movement can occur at any time and in any sequence.

With the possible exception of facial animation, the deformations required in 3-D character animation are controlled with an array of influences called **bones**. These bones are embedded within the portion of the character that the animator wants to move. Bones are usually embedded in chains of two or more, because obviously one bone would do nothing more to a character than a simple Move command. Chains of bones are placed in a specific order that creates a **hierarchy**. The first bone, by default, is assigned the role of **parent**, whereas the second bone becomes the **child** of the first. If a third bone is added, it becomes the child of the second and so forth. The hierarchy is necessary and useful because eventually you will want to create realistic skeletal motion using the bones, and the hierarchy determines how the bones will interact with each other.

For instance, when building a human character, animators often begin with character's spine, building from the top down. At the hip, two chains of bones branch out downward to control the feet. The topmost bones of the legs (you might call them lthigh and rthigh) are the parent bones for each leg. When you move these bones, the rest of the leg bones (rshin, rfoot, rthumbtoe, rindextoe, and so on) all move along proportionately. That isn't to say that they will naturally place themselves like the bones of a real human (wouldn't that be a nice feature?), but if you create the bones in the proper order, at least you won't have to move each one individually (see Figure 7-13).

Working from the top of the spine, you would add arm bones on each side by forming one chain first, then **mirroring** or **cloning** that chain, and attaching it to the other side of the spine. Finally, a neck bone, extending upward from the spine, and a head bone would finish off the simple human skeleton. However, other considerations must be made when we get to the head (you might also want to take note that a tail, if needed, would also require a set of bones). If you wanted a mass of hair (or a jester's cap) to move about on the head, another bone or bones could be added to the head.

Figure 7-13

A perspective view in Maya showing a human form with the appropriate bones and joints placed for the legs and feet. Note that Maya, in addition to bones, offers the capability to place three types of joints.

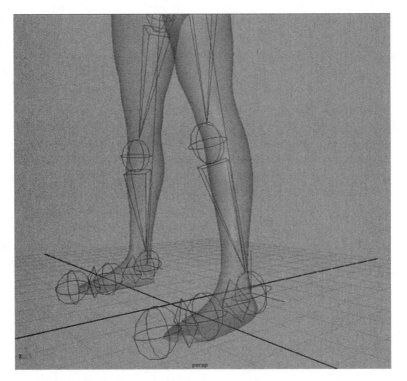

Constraints and Joint Types

Computers have no way of knowing a bone you have placed in a character represents a real bone in a real limb, and computer bones have to be told they are bones. In other words, once you've linked up some bones, you have to tell the computer which rules to follow with regard to their motion.

For instance, if you have a knee joint between two bones, how would the computer know that the knee can only rotate on one axis and that it can't rotate more than 180 degrees (or even less, if you figure the space taken up by muscles)? You do this with **constraints** (in LightWave this is called the **range,** which has a minimum and maximum). Constraints are assigned to the relationships between bones to make them act more like real bones and joints. Assigning constraints that emulate real creatures simplifies your animation control process and often speeds up your calculations.

Maya, incidentally, offers the advantage of joints between bones. Three types of joints are allowed. A ball joint can rotate in all three axes like the joint between the atlas (or topmost) vertebra and the skull. A universal joint can rotate in two axes like an ankle joint. A hinge joint can rotate in only one axis like the knee joint. By using these specific joints, Maya users can quickly build skeletons with constraints already started.

Kinematics: Forward and Inverse

Kinematics refers to the method by which the computer simplifies the calculations required to move your character's limbs, which is a process that would otherwise be difficult to manage. Two types of kinematics exist: *forward kinematics* (FK) and *inverse kinematics* (IK).

FK has been around the longest and is now incorporated so well into animation programs that it is virtually a part of the bone-creating process. If you succeed in creating and placing the bones into your character in the correct order, you've no doubt created a perfectly workable FK chain.

Take one of these chains, such as the three bones representing the two long bones of the arm and one bone for the hand. If you wanted to create realistic arm swinging as a character walks without kinematics, you would simply create a keyframe with the shoulder back, the elbow joint cocked back, and the wrist pulled forward and up. Then you would create another frame with the shoulder forward, the elbow almost locked in line with the body, and the wrist almost straight below the elbow. To save some time, you might create a third keyframe the same distance from the second keyframe as the second is from the first, and then copy the bone position attributes from the first keyframe into the attributes for the third keyframe.

When animating this sequence, the top bone in the FK chain, the shoulder, would swing forward, and all the other bones would follow along likewise, limited only by the positional attributes that you had stored in the second and final keyframes. These attributes act as temporary restraints, but otherwise all the bones in the arm are importing their motion (or kinetics) from their parent bones. This is FK.

You might use the same technique to create the effect of a boxer getting ready to throw a punch by thrusting his right shoulder forward in preparation to launch the right fist at his opponent's glass jaw. As the shoulder thrusts forward, the rest of the bones in the arm move. You don't need to position them because FK does the work.

However, when it comes time to throw the punch and hit a target, FK would be a cumbersome way to work, because you would be attempting to aim the fist with the shoulder. Mike Tyson personally advised me against this.

In order to achieve goal-directed bone motion, you need IK. This requires a bit more math than FK. The mathematical process begins when you position the end of the hand, using the last bone (the term is **gnomon**, though **goalnull** and **end effector** are also used). The computer examines all the positional attributes of the bones and joints from the gnomon to the top rotation axis of the top bone. It then calculates a solution (Maya and 3ds max call this the **IK solver**) that results in the correct placement of all the bones in the chain.

In other words, once FK has thrust the shoulder of the boxer forward, IK enables you to place the closed fist, knuckles first, against the hapless chin of the victim. Press a button and see the entire arm arrange itself perfectly for the knockout!

Another great example of IK is when you have a rotating handle like that found on a coffee grinder. The simplest way of arranging a character to turn this handle is to first animate the rotation of the handle and then constrain or parent the hand of the character to the handle using IK solvers. Presto, with very little work, the computer figures out all the bones of the character from the gnomon at the hand all the way up to the arm and through the spine.

Facial Considerations

Although an intricate assortment of bones can be designed to control facial expressions, which are important in establishing a character's personality, creating facial animation can be achieved in other ways as well.

Many animators prefer to use **shape** or **morphing animation** to exert the maximum control on a face. This form of animation requires that you deform the face by manipulating the polygons, splines, or control vertices (depending on which type of object you are using) until you attain a particular expression, which would be entered as a keyframe. Then you would manipulate the face into a different expression and save that as a second keyframe. To create the facial change, you would indicate a morph transformation between the two keyframes.

Obviously, it is wise to start with a natural expression and make each key facial expression depart from that. By running the sequence backward, a return to the natural expression is easily created.

Most animation programs offer a means by which audio files of the character's voice can be inserted into the animation timeline. By doing this, you can quickly assemble all the essential facial expressions required for **lip synching**.

Programs also enable a combination of skeletal and morph animation. In such cases, you might want to create bones for the upper and lower jaw, and perhaps one for the tongue, in order to control mouth movements, whereas morphing is used to control finer touches. Of course, the jaw and tongue bones should be parented to the head bone.

Animating a Character

Now comes the most fascinating part of character animation: manipulating the character to produce a sequence of frames. This is done exactly the same way as all other animations discussed thus far: by moving the various parts of the character and saving each pose as a keyframe on the timeline.

But instead of flying logos or dividing molecules, character animation evokes the subtle aspects of real-world phenomenon with which most audiences are familiar. Everyone knows what a walk looks like, and although everyone has a different type of walk, a bad one sticks out like a sore thumb.

I would like to take a moment to pay homage and also to recommend a superb book for the study of character animation (and for many other animation techniques as well). I would be a fool's dumb brother to try and improve on the work of Richard Williams's book, *The Animator's Survival Kit: A Manual of Methods, Principles, and Formulas for Classical, Computer, Games, Stop Motion, and Internet Animators*. Although Rick is a traditional cell animator (best known for his masterpiece creation, *Who Framed Roger Rabbit*), the techniques he reveals for character animation are precisely applicable to everything you will do in 3-D.

Rick believes, and I concur, that today's animation artists do not spend enough time practicing their basic life sketching and other traditional drawing skills. Although many art schools and colleges have dropped life drawing from their curricula, this is a skill you can easily incorporate in your daily routine by simply carrying a sketchbook and charcoal pencil wherever you go.

Sliders

Many animation programs enable you to finish off the rigging of a character by assigning any of the various channel controls to a graphical slider (see Figure 7-14). Therefore, instead of moving the leg gnomon up and down, for instance, you can move a slider named right leg. By combining these controls into one slider, you can build fairly complex combinations

Figure 7-14
A character face in 3ds max rigged to a set of sliders (shown in columns to the left and right of the model) that control various expressions. The cursor is moving Slider 05 to angle the right brow downward.

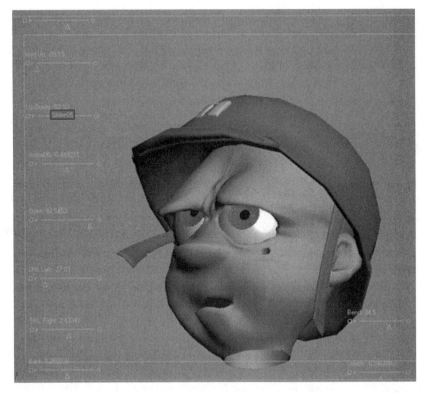

that result in a control interface, which is fluid and useful for fast positioning a character, such as a stringed puppet.

I recall how master animator Patrick Starace once figured out how to take a simple *musical instrument digital interface* (MIDI) controller and, using a *universal serial bus* (USB) interface, connect it to the slider controls of his Softimage installation. The MIDI controller cost about $120 and offered 12 sliders that could be reconfigured to any of this character's controls. Twelve controls is quite a lot for a character like the lotus-posed, chip-eating Buddha that Pat created for a TV commercial and that was completely operated by the inexpensive MIDI panel.

Replicate and Exaggerate

Your skill in replicating the realism of the world around you or exaggerating that realism to a controllable degree will make or break you as a

3-D animator. Realism is often necessary to create believable character motion. The pose of a tired old man sitting on a subway bench that you sketch one Sunday morning may become the pose of a tuckered-out super-hero after a long bout with evildoers. With practice, you can begin to build a library of reliable images and develop the technical skill to replicate them, but this skill will never develop if you don't practice sketching. If all you know is the Wacom tablet and punching CVs to match someone else's drawing, you will never fulfill your destiny as a creator. You'll get stuck in an assistant-level role, forever doomed to serve someone else's superior creative skill.

Although sketching is the basic art form of the animator, sculpture is also essential in understanding the conversion of a 2-D creation into a plastic, 3-D environment. Buy a big lump of clay and keep it handy. Use it to make a face or model a toy. Take a portrait you sketched and sculpt the face in three dimensions.

Sketching and clay modeling are two of the most inexpensive methods for developing skills that will prove highly valuable when you sit down at a $10,000 workstation. You may not be able to afford the workstation or be able to practice at one on a regular basis, but whenever you have a spare moment, you can sketch and model.

Using an inexpensive video camera is also a good way to capture the sub-tleties of the real world around you. A few years ago, we had to model a real-istic cheetah for a commercial. Not content with the stock footage we located on the Internet, the animator went to the Bronx Zoo and videotaped an hour of cheetahs pacing and running through their compound. We then captured the best clips on an NLE system and imported the keyframes into 3ds max. There the animator took the still frames and matched up the wire-frame model of the cheetah we built earlier. This produced authentic results and since then this has been our modus operandi for replicating nature.

But the client hated the results! Although the cheetah was anatomically precise (it even fooled a zoologist who came to see our work), the client thought the cheetah looked "stiff legged." Well, that's the way cheetahs really walk. "No, make it more fluid. Make it like a cat." So we got some cat walks on video and merged the two sets of points. The client liked that bet-ter but was still not completely satisfied.

Apparently, the replication of reality is not always what people expect to see in animation. What they really want is creative exaggeration. So we went back to our cheetah/tabby cat and pushed and pulled the bones a bit this way and a bit that way until it sort of slinked across the screen. It slinked in a way that no real creature slinks. But guess what? The client loved it and after three months of fiddling (and waiting for our final payment) we satisfied the client and rendered for the last time. It was an exaggeration of reality that the client wanted, not replication.

So when you are faced with animating a character and you've studied all the Eadweard Muybridge[2] photos, stock footage, and anatomical drawings, feel free to add a bit of your own personality to the project. That's when you elevate yourself from the level of mechanic to the level of artist.

Function Curves

After you have test-rendered your character animation, the results might not be what you expected. My experience with novice animators (and not-so-novice clients and producers) is that any initial attempts to create smooth movement using keyframes often result in unaesthetic "cornering." In other words, you may want a smooth sweep of a hand from point A to point C, with the midpoint at point B. Instead of a smooth continuous flow from A through C, you get a noticeable effect at point B, which animators call a "corner," because the most common application of three-point keyframes is a flying object, and when the object badly transitions through point B, it seems to turn a visible corner in space. Actually, this is a good thing, because it proves that computers still aren't clever enough to think like humans (and the day they are, *all* our jobs are toast).

If you get results that don't look as smooth or as realistic as you'd like, you will need to use the tools that control your keyframes. In 3dx max, these are called **function curves**, in LightWave they're called **motion graphs**, and in Maya they are called **graph editor animation curves**.

These functions all work basically the same. These graphical controls enable you to view, often during playback, the actual characteristics of the animation as they change over time. These controls appear as various graphed lines and curves. Therefore, if you see cornering in an animation, it will appear as a sharp point between two lines.

---

[2]To win a bet, the governor of California hired Muybridge (1830–1904), a photographer, to develop a means of photographing horses while in movement. Muybridge's sequential photos of a horse galloping won the governor's bet and Muybridge's fame. Today you can find his public domain sequences on the Internet. They are useful in animation motion studies of almost any animal, including humans.

The function curve, motion graph, and animation curve tools enable you to modify the graphs with editable splines. Just as spline curves enable the creation of smooth surfaces on a 3-D object, they also serve to provide smooth transitions between keyframes in an animation. By altering the shape and steepness of the curve, you can alter the appearance of your animation. These functions therefore provide a valuable interactive feedback mechanism you can use to tweak your animations to perfection.

INDEX